Wolfgang Fischer

Design of compact climbing robots for power plant inspection

Wolfgang Fischer

Design of compact climbing robots for power plant inspection

Innovative mechanisms and vehicle structures for moving in generators, steam-chests or other complex-shaped environments

Südwestdeutscher Verlag für Hochschulschriften

Impressum/Imprint (nur für Deutschland/only for Germany)
Bibliografische Information der Deutschen Nationalbibliothek: Die Deutsche Nationalbibliothek verzeichnet diese Publikation in der Deutschen Nationalbibliografie; detaillierte bibliografische Daten sind im Internet über http://dnb.d-nb.de abrufbar.
Alle in diesem Buch genannten Marken und Produktnamen unterliegen warenzeichen-, marken- oder patentrechtlichem Schutz bzw. sind Warenzeichen oder eingetragene Warenzeichen der jeweiligen Inhaber. Die Wiedergabe von Marken, Produktnamen, Gebrauchsnamen, Handelsnamen, Warenbezeichnungen u.s.w. in diesem Werk berechtigt auch ohne besondere Kennzeichnung nicht zu der Annahme, dass solche Namen im Sinne der Warenzeichen- und Markenschutzgesetzgebung als frei zu betrachten wären und daher von jedermann benutzt werden dürften.

Coverbild: www.ingimage.com

Verlag: Südwestdeutscher Verlag für Hochschulschriften GmbH & Co. KG
Dudweiler Landstr. 99, 66123 Saarbrücken, Deutschland
Telefon +49 681 37 20 271-1, Telefax +49 681 37 20 271-0
Email: info@svh-verlag.de

Zugl.: Zürich, ETH, Diss. Nr. 18975, 2010

Herstellung in Deutschland:
Schaltungsdienst Lange o.H.G., Berlin
Books on Demand GmbH, Norderstedt
Reha GmbH, Saarbrücken
Amazon Distribution GmbH, Leipzig
ISBN: 978-3-8381-2770-5

Imprint (only for USA, GB)
Bibliographic information published by the Deutsche Nationalbibliothek: The Deutsche Nationalbibliothek lists this publication in the Deutsche Nationalbibliografie; detailed bibliographic data are available in the Internet at http://dnb.d-nb.de.
Any brand names and product names mentioned in this book are subject to trademark, brand or patent protection and are trademarks or registered trademarks of their respective holders. The use of brand names, product names, common names, trade names, product descriptions etc. even without a particular marking in this works is in no way to be construed to mean that such names may be regarded as unrestricted in respect of trademark and brand protection legislation and could thus be used by anyone.

Cover image: www.ingimage.com

Publisher: Südwestdeutscher Verlag für Hochschulschriften GmbH & Co. KG
Dudweiler Landstr. 99, 66123 Saarbrücken, Germany
Phone +49 681 37 20 271-1, Fax +49 681 37 20 271-0
Email: info@svh-verlag.de

Printed in the U.S.A.
Printed in the U.K. by (see last page)
ISBN: 978-3-8381-2770-5

Copyright © 2011 by the author and Südwestdeutscher Verlag für Hochschulschriften GmbH & Co. KG and licensors
All rights reserved. Saarbrücken 2011

Abstract

Power plants and similar facilities have to be inspected regularly for maintenance and security reasons. Automating this task by using compact climbing robots can not only decrease the total inspection time, but even allows for reaching areas that were inaccessible with conventional methods. However, the number of successful industrial applications is still quite low. For this reason, the main motivation of this thesis is to deeper analyze the general applicability of this technology in the field of power plant inspection; to describe, classify and compare the most recent innovations in this field; and show their application in real business cases.

In the introduction chapter, the main functions and components of a power plant are explained, pointing to the importance of regular inspections and the advantages that can be achieved by using compact climbing robots for this task. This introduction is followed by an overlook on typical application scenarios, the most important market requirements and their influence on the functions of a compact climbing robot – concluding to the central goal within this research – to find new vehicle structures that combine high mobility and low complexity.

A general classification for most basic types of compact climbing robots is provided in the third chapter, which is structured according to the two main functions for climbing – adhesion and locomotion. It concludes with a comparison of the most common vehicle structures, which is mainly based on the business relevance in the field of power plant inspection and the relative performance that can be achieved with the most recent prototypes. According to this comparison, robots with roll-legged locomotion, passive adhesion and mechanisms that increase the mobility on specific obstacles seem to be the most promising alternative for most applications.

For this reason, the next chapter mainly deals with this type, sets the main focus on mechanisms for increasing the mobility on different types of obstacles and provides a detailed description, comparison and performance-evaluation for the most important mechanisms and vehicle concepts. The methodology of this performance-evaluation is mainly based on tests with real prototypes – or in the case of external developments also on literature, video-analysis and discussion with the researchers. In some cases, also mechanical calculation models have been established. Given the huge number of analyzed prototypes (30) and environment challenges (43), this chapter only provides the description of the basic concepts and the rough comparison of prototypes according to their performance on the environment challenges that were taken from real power plant applications. Most technical details about prototype design, calculation models and tests can be found in the referred papers – in the case of prototypes with mayor importance, these details are also included in the corresponding case study, which can be found in the last part of this thesis.

These case-studies show the complete development process in three typical projects on inspection robots – gas tanks, steam chests and generator housings. They start with a detailed analysis of the applications - pointing both to the business-relevance for the customer and to the main environment challenges that need to be solved with innovative robot design. This specification analysis is followed by a description of the different prototype generations, focusing on the core advantages towards previous designs and how the most difficult design challenges have been solved. Each case study concludes with a summary of the core innovations, a short outlook and some "lessons-learned" for future projects. Four other projects that showed other types of challenges than the main ones, but achieved important economic and/or scientific success are summarized briefly in the last section of this chapter – generator air gaps, flying and docking in boilers, turbine inspection and power lines.

The main contributions of this work are first the description, analysis and discussion of more than 30 innovative obstacle-passing-mechanisms and vehicle structures that significantly increase the mobility of compact climbing robots. More than ten of them were developed in the context of this thesis. In addition to these descriptions, these mechanisms are structured, classified and compared among each other and towards previous technologies. Finally, this thesis shows how these solutions can successfully be applied in real power plant applications.

Zusammenfassung

Kraftwerke und ähnliche Anlagen müssen regelmässig aus Zuverlässigkeits- und Sicherheitsgründen inspiziert werden. Die Automatisierung dieser Aufgabe mithilfe von kompakten Kletter-Robotern reduziert nicht nur die Gesamt-Inspektionszeit, sondern erlaubt sogar den Zugang zu Zonen, welche mit konventionellen Methoden nicht erreicht werden können. Trotz dieser Vorteile ist die Anzahl an industriellen Anwendungen noch relativ gering. Aus diesem Grund besteht die Hauptmotivation zu dieser Arbeit darin, die generelle Nutzbarkeit dieser Technologie im Bereich der Kraftwerksinspektion zu analysieren, die aktuellsten Innovationen in diesem Forschungsfeld zu beschreiben, zu klassifizieren und zu vergleichen; und deren Anwendung anhand von realen Fallbeispielen aus dem industriellen Alltag zu erläutern.

Hierbei werden im Einleitungs-Kapitel die Hauptfunktionen und Komponenten eines Kraftwerks erklärt, wobei besonders darauf eingegangen wird, wie wichtig regelmässige Inspektionen sind und welche Vorteile hierbei mit kompakten Kletterrobotern erzielt werden können. Dieser Einleitung schliesst sich ein Überblick auf typische industrielle Anwendungen an, gefolgt von den wichtigsten Markt-Anforderungen und deren Einfluss auf die Funktionen eines kompakten Mobilroboters mit vertikaler Mobilität. Hieraus wird auf die zentralen Ziele innerhalb dieser Forschung geschlussfolgert – neue Fahrzeug- und Antriebs-Strukturen zu finden, welche eine hohe Mobilität bei geringer Komplexität ermöglichen

Eine umfassende Klassifizierung für die grundlegenden Bauformen bei kompakten Kletterrobotern wird im dritten Kapitel präsentiert. Die Struktur orientiert sich hierbei an den beiden Hauptfunktionen für das Klettern – Haftung und Fortbewegung. Am Schluss dieses Kapitels steht ein Vergleich von aktuellen Fahrzeugstrukturen und Prototypen, welcher hauptsächlich auf Kundenanforderungen im Bereich der Kraftwerksinspektion basiert, sowie auf der relativen Leistungsfähigkeit, die mit den neuesten Prototypen erreicht wird. Gemäss diesem Vergleich erscheinen Roboter mit rollförmiger Fortbewegung (Räder, Raupenketten oder felgenlose Räder/Whegs), passivem Haftprinzip und zusätzlichen Mechanismen zur Überwindung von spezifischen Hindernissen als am aussichtsreichsten für die meisten Anwendungen.

Aus diesem Grund befasst sich das nächste Kapitel hauptsächlich mit dieser Gruppe, setzt den Fokus auf Mechanismen zur Erhöhung der Mobilität an verschiedenen Arten von Hindernissen und liefert detaillierte Beschreibungen, Vergleiche und Leistungsbewertungen für die wichtigsten Mechanismen und Fahrzeug-Konzepte. Die Methodik dieser Leistungsbewertung basiert hauptsächlich auf Versuchen mit realen Prototypen – oder im Fall von externen Entwicklungen manchmal auch auf Literaturrecherche, Videoanalyse und/oder Diskussion mit den Forschern. In einigen Fällen wurden auch mechanische Berechnungsmodelle erstellt. In Anbetracht der grossen Anzahl an analysierten Prototypen (30) und Kundenanforderungen (43) liefert dieses Kapitel allerdings nur die Beschreibung der grundlegenden Konzepte und den groben Vergleich der Prototypen anhand ihrer Leistungsfähigkeit in Bezug auf die Kundenanforderungen aus den analysierten Industrie-Anwendungen. Die technischen Details zur Auslegung einzelner Komponenten, die Berechnungsmodelle und die Dokumentation der Experimente können in den zitierten wissenschaftlichen Publikationen nachgeschlagen werden – bei den Prototypen mit besonders hoher industrieller Relevanz sind die wichtigsten technischen Details auch in der entsprechenden Fallstudie beschrieben, welche sich im letzten Kapitel dieser Arbeit finden.

Diese Fallstudien zeigen den vollständigen Entwicklungsprozess in drei repräsentativen Projekten zu Inspektions-Robotern – Gas-Tanks, Einströmkästen vor Dampfturbinen (steam chests) und Gehäuse von Generatoren. Sie starten jeweils mit einer detaillierten Analyse der Anwendung, wobei sowohl auf die wirtschaftlichen Vorteile für den Kunden als auch auf die wichtigsten technischen Herausforderungen eingegangen wird, welche mit innovativem Roboter-Design gelöst werden sollen. Dieser Diskussion der Spezifikationen folgt eine Beschreibung der verschiedenen Prototyp-Generationen, wobei besonderer Wert auf die zentralen Vorteile gegenüber bisherigen Bauformen gelegt wird, sowie darauf, wie bestimmte Detail-Probleme im Bereich des Antriebsstrangs gelöst wurden. Jede Fallstudie endet mit einer Zusammenfassung über die zentralen Innovationen, liefert einen Ausblick auf nachfolgende Forschungstätigkeiten und legt dar, was für zukünftige ähnliche Projekte gelernt werden kann. Zusätzlich zu den drei Haupt-Projekten werden auch noch vier weitere kurz erwähnt, bei denen besondere Erfolge in der wirtschaftlichen Umsetzungen und bei den wissenschaftlichen Publikationen erzielt wurden – auch wenn die technischen Herausforderungen und Innovationen einen weniger starken Bezug zu den zentralen Forschungszielen im vorherigen Kapitel haben. Diese Projekte behandeln innovative Roboter für die Inspektion von Generatoren, Kohle-Brennkammern, Gas-Turbinen und Hochspannungsleitungen.

Der zentrale wissenschaftliche Beitrag dieser Arbeit besteht in der Beschreibung, der Analyse und der detaillierten Diskussion von über 30 innovativen Mechanismen zur Überwindung von Hindernissen, sowie Fahrzeugstrukturen die signifikant zur Erhöhung der Mobilität von kompakten Kletterrobotern beitragen. Mehr als zehn davon wurden im Zusammenhang dieser Arbeit neu entwickelt. Zusätzlich zu dieser Beschreibung werden diese Mechanismen auch das erste Mal strukturiert dargestellt, klassifiziert und verglichen – sowohl untereinander als auch im Vergleich zu vorherigen Technologien. Des Weiteren wird gezeigt, wie diese Lösungen erfolgreich in realen Industrie-Anwendungen im Bereich Kraftwerks-Inspektion eingesetzt werden können.

1 INTRODUCTION TO POWER PLANT INSPECTION 1

1.1 Functionality of power plants and main types 1
   1.1.1 Thermal power plants with a steam cycle 1
   1.1.2 Gas-turbines and combined-cycle power plants 2
   1.1.3 Power plants using renewable energies 2
1.2 Main components of a power plant 3
   1.2.1 Generator 4
   1.2.2 Turbines 5
   1.2.3 Steam cycle with condenser, pump, pipes and tubes 6
   1.2.4 Reactors, boilers and exhaust purification systems 7
   1.2.5 Components outside the power plant 8
1.3 Inspection of the core components 8
   1.3.1 Need for regular inspections 8
   1.3.2 Economical impact 9
   1.3.3 Limitations of conventional inspection methods 10
   1.3.4 Advantages of inspections with mobile robots 11
   1.3.5 Focus on compact climbing robots 11

2 REQUIREMENTS AND FUNCTIONS 13

2.1 Environment restrictions 13
   2.1.1 Overview on typical environments in power plants 13
   2.1.2 Size restrictions and height of vertical sections 18
   2.1.3 Typical obstacles 18
   2.1.4 Surface characteristics and other hazards 19
2.2 Market requirements and additional needs 19
   2.2.1 Payload 19
   2.2.2 Safety, reliability and robustness 21
   2.2.3 Universality and modularity 21
2.3 Main functions of a compact climbing robot 22
   2.3.1 Climbing with high mobility 22
   2.3.2 Other main functions in an inspection robot 23
2.4 Main goals for the research 24

3 CLASSIFICATION OF CLIMBING ROBOTS 25

3.1 Adhesion principles 25
   3.1.1 Mechanical adhesion 26
   3.1.2 Ferromagnetic adhesion 31
   3.1.3 Alternative adhesion principles 36
   3.1.4 Active vs. passive principles 39
   3.1.5 Comparison and performance evaluation 40
3.2 Locomotion 41
   3.2.1 Roll-legged locomotion 42
   3.2.2 Swing-legged locomotion 44
   3.2.3 Hybrid locomotion 45
   3.2.4 Relative performance in mobility and complexity 46
3.3 Overview and performance comparison 50
3.4 Conclusion and focus of the next chapters 52

4 OBSTACLE-PASSING WITH COMPACT CLIMBING ROBOTS 53

4.1 Detailed classification of obstacles 53
   4.1.1 Inner transitions (concave corners) 53
   4.1.2 Outer transitions (convex edges) 57
   4.1.3 Curvatures 63
   4.1.4 Combinations 65
   4.1.5 Other hazards 65
   4.1.6 Influence of the obstacle size relative to the robot 66
   4.1.7 Mobility estimation for climbing robots 67
4.2 Mechanisms for increasing the robot mobility 69
   4.2.1 Mechanisms for inner transitions (concave corners) 69
   4.2.2 Mechanisms for outer transitions + combined obstacles 83
   4.2.3 Other mechanisms for increasing the mobility 90
4.3 Robotic vehicles and their performance 97
   4.3.1 Classification based on the obstacle-passing-mechanisms 97
   4.3.2 Brief description of the most relevant vehicles 99
   4.3.3 Comparison 104
4.4 Summary and conclusion 106

5 CASE STUDIES 107

5.1 Gas storage tanks 108
5.1.1 Motivation and business relevance 108
5.1.2 Environment specifications 108
5.1.3 Prototypes 110
5.1.4 Innovation, outlook and lessons learned 115

5.2 Steam chests 117
5.2.1 Motivation and business relevance 117
5.2.2 Environment specifications 118
5.2.3 Prototypes 119
5.2.4 Innovations, outlook and lessons learned 132

5.3 Generator back housing 135
5.3.1 Motivation and business relevance 135
5.3.2 Environment specifications 135
5.3.3 Prototypes 138
5.3.4 Innovation, outlook and lessons learned 149

5.4 Successful projects with other technical challenges 151
5.4.1 Generator air gap 151
5.4.2 Lightweight helicopter dock 154
5.4.3 Micro-crawlers for boiler tubes and steam turbines 155
5.4.4 Cable crawler for power line inspection 156
5.4.5 Innovation, outlook and lessons learned 157

6 SUMMARY AND CORE VALUE OF THIS WORK 159

APPENDIX 1: COMPARISON MATRIX 163
APPENDIX 2: REFERENCE LIST 168

1 Introduction to power plant inspection

Power plants and similar facilities have to be inspected regularly for maintenance and security reasons. Automating this task using mobile robots can decrease the total inspection time and even allow for reaching areas that were inaccessible with conventional methods. Thus, there is a great potential for both increasing the inspection quality and simultaneously decreasing inspection costs and time. As the business of inspecting power plants is very profitable, new innovations in this field bring significant economical benefits.

In the first section, the most important types of power plants are presented briefly, with the main focus on how the energy is transformed from one form to another; and which components are necessary for these processes. The next section shows some more details about the most important components, how they work, what the most important design challenges are and what the differences within one group of components are. The need for inspecting these components regularly; and the economical benefits that can be generated by using compact climbing robots for this task is explained in the last section of the introductory chapter – leading over to the main motivation of this work:

> Make power plant inspection simpler, faster and cheaper.

1.1 Functionality of power plants and main types

This chapter provides a brief overview how electric energy is generated in a power plant (often also called "power station", mainly in UK), and which processes and components are necessary for transforming the energy coming from natural resources to the electrical energy for the end user.

1.1.1 Thermal power plants with a steam cycle

The most common type of power plant is the thermal power plants with steam cycle. In this type, fuel is burned to produce heat which can then be exploited in a thermodynamic cycle to generate dynamic pressure of hot steam. This dynamic pressure is then transformed by a steam turbine into kinetic rotational energy of a shaft. A generator finally converts this rotational energy into electricity. Four main categories in this type of power plant exist: Nuclear, coal, biomass, waste.

1.1.2 Gas-turbines and combined-cycle power plants

A very special component within the thermal power plants is the gas-turbine. There, the combustion process takes place in a way that it directly generates the dynamic pressure that is necessary for rotating the turbine blades. To improve the overall efficiency, also the heat from the exhaust gas can be used to power a steam-cycle – similar as in other thermal power plants. These so-called "combined-cycle-power-plants" (CCPP) can reach efficiencies up to 60%, which is significantly higher than the efficiency of a simple gas-turbine power plant (35-40%) or other non-nuclear power plants (Coal, biomass, waste: 40-45%).

But also "simple" gas-turbine power plants without steam cycle exist. These power plants can be started at very short time and thus are often used only for the peak demand where the relatively low efficiency is not so crucial. They also require less capital investment than other types. For this reason, they are mainly installed in regions where there are no possibilities to cover the peak demand with pumped storage hydropower and/or in countries with weak economy (no money to invest in CCPPs or nuclear power plants).

1.1.3 Power plants using renewable energies

Electric energy can also be generated with other means instead of burning fuel. Currently, power plants based on renewable energies still generate less than 5% of the globally installed electrical power, but their significance will grow in the next decades due to the shortage of fossil fuels and uranium. For this reason, their processes and components are also analyzed in this work.

The most frequent type of power plant that uses renewable energies is hydroelectricity. In these power plants, the potential energy of dammed water is driving a turbine and generator. Other types of hydro-electric power plants are installations that use the kinetic energy of marine currents, tidal streams or waves.

For exploiting the energetic potential of wind, several parks of wind turbines have been installed during the last years. In many European countries, the installed power of wind parks is already bigger than the one with hydroelectricity; and the rate of economic growth within this type is the biggest among all main types of power plants. A typical wind turbine is normally formed by a tubular steel tower (60-90m height), three blades of 20-40m length, and a generator. As the blades have a higher efficiency when turning slowly, normally a gear box is mounted before the generator, to speed up to the normal line frequency of 50Hz.

Also the heat or the radiation of the sun can be directly exploited for generating electrical energy. While solar-thermal power plants concentrate sunbeams with parabolic collectors to a focal point and then exploit the heat difference towards the environment in a steam cycle, solar cells directly transform the sunlight into electricity by using the photovoltaic principle.

Burning bio mass or waste for heating and producing electricity in a steam cycle is another form of renewable energy. The structure and the components of these installations are very similar to coal power plants. Another possibility for powering a steam cycle by renewable energies is geothermal heat. Power plants of this type are mainly installed in regions with high volcanic activity (e.g. Iceland).

1.2 Main components of a power plant

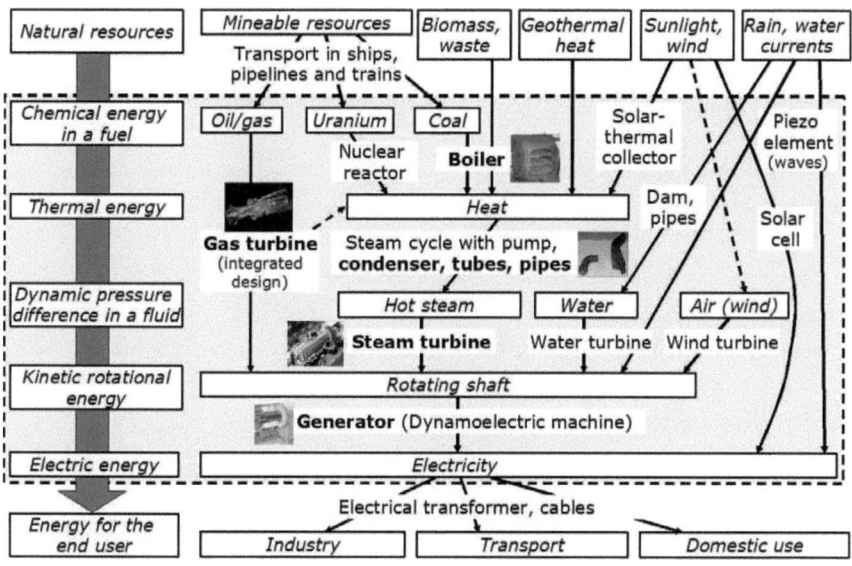

Fig. 1-1: Energy conversion in power plants and main components (Bold and with picture: Components within the core business of ALSTOM-Power-Systems, analyzed more detailed in this work)

As already stated in the last section, most types of power plants (except some "exotic" ones such as solar cells or piezo arrays in water) generate electricity with the combination of a turbine and a generator. In thermal power plants, also a boiler, a nuclear-reactor or a gas turbine, plus the installations for a steam cycle are necessary. The main forms of energy, the components necessary for transforming from one form to another, and in bold letters the components that are within the core business of ALSTOM-power as original equipment manufacturer (OEM) can be seen in Fig. 1-1. As this project was mainly funded by ALSTOM power, these components have higher priority than the others and thus have been analyzed more detailed.

The following section shows some how these components work, what the most important design challenges are and what the differences within one group of components are.

1.2.1 Generator

For transforming the kinetic energy of the rotating shaft into electric energy, the principle of electromagnetic induction is used. This is done in a generator, also often called dynamoelectric machine. In such a generator, the rotation of the electromagnet in the rotor creates a rotating magnetic field. This rotating field then induces a current in the induction coils of the stator. To avoid losses coming from unwanted eddy-currents, the stator is designed out of several small parts of steel, also called "teeth". Most generators in power plants are realized as synchronous machines, with the electromagnetic coils in the rotor excited by an external source of continuous current. This type – compared to asynchronous machines – is more difficult to control and more expensive regarding the materials, but has a better efficiency and also guarantees a stable operation when the consumption coming from the electrical grid is relatively unsteady.

As even in this design some losses cannot be avoided, the need for cooling becomes more and more important the bigger the size of a generator is. While for generators below 200MW (small power plants) air-cooling is sufficient, the bigger generators in coal and nuclear power plants are normally cooled with water or hydrogen – with both systems requiring quite complex networks of tubes and heat exchangers.

Regarding the orientation in respect to gravity, most generators are mounted with the axis horizontally. The only exception is in hydroelectric power plants with Francis or Kaplan turbines (Fig. 1-3-e), requiring a vertical position of the axis.

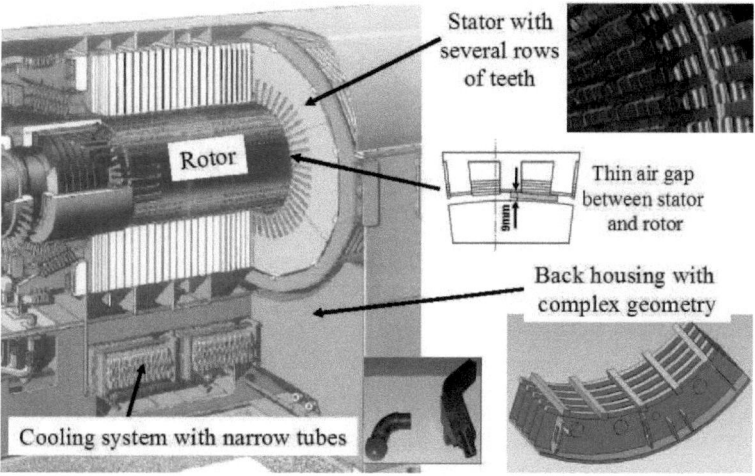

Fig. 1-2: Cut-view of a large generator in power plants (~500MW), with its main areas that are interesting for being inspected with small robots

1.2.2 Turbines

To make the generator shaft rotate, the dynamic pressure of a fluid is exploited. This is normally done in a turbine. In contrast to reciprocating machines (e.g. diesel engines or the very first steam engines), the motion is continuous which allows for less accelerated mass and thus for building relatively compact still at very high power. For this reason, the turbine principle is used in almost all types of power plants.

As the overall design and the blade geometry mainly depends on the operating fluid and the pressure difference, the most common classification of turbines is according to the fluid that passes through it: Gas-, steam-, water- or wind-turbines. Within the water turbines, there is also a sub-classification according to the working principle: Pelton-, Kaplan- and Francis-turbines. While the Pelton-type is normally used for large heads (50-1300m), for small heads Francis- or Kaplan-turbines usually result in better performance.

Fig. 1-3: Different types of turbines (a) Steam turbine (b) Gas turbine GT26 (both: ALSTOM-Power [110]) (c) Wind turbine in Hannover-Kronsberg [111] (d) Pelton turbine from Blue Water Power [112] (e) Kaplan turbine at "Erlebniskraftwerk Walchensee" [113]

As already mentioned in 1.1.2, in gas turbines (Fig. 1-3, c) the turbine also includes a combustion chamber and a compressor. For this reason, it can both exploit the dynamic pressure generated in the combustion process to drive the gas turbine blades and the hot temperature to power a steam cycle.

1.2.3 Steam cycle with condenser, pump, pipes and tubes

To transform the thermal energy coming from the boiler into the dynamic fluid pressure that is then exploited by the turbine, a closed-loop steam-cycle is normally applied: Water at high pressure gets evaporated with the heat of the boiler, to produce dry saturated steam at hot temperature. This saturated steam is then expanded in the turbine, where the kinetic energy gets extracted. To close the cycle, the wet vapor coming from the turbine enters a condenser where it is condensed at a constant pressure and temperature to become again saturated water (at low pressure). To keep the condenser at constant temperature, water (from outside) is evaporated in cooling towers. If the plant is close to a big city, the heat from the condenser is normally used for district-heating. Before the water again enters the boiler, its pressure gets increased by a pump. The energy consumption of the pump is relatively small compared to the boiler.

1. Introduction to power plant inspection.

For connecting boiler, turbine, condenser and pump, a system of tubes and pipes is necessary. These pipes are normally delivered by the business unit that manufactures the boiler, except the last distributor parts before the steam turbine (steam chests) that are still considered to be part of the turbine.

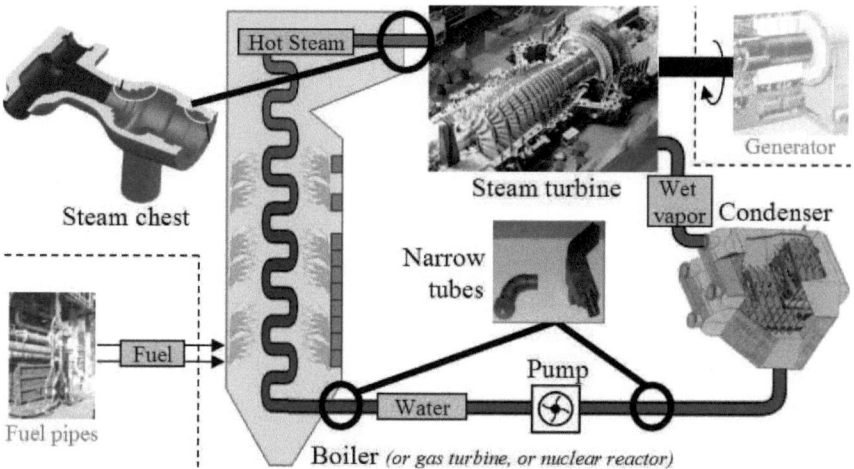

Fig. 1-4: Steam cycle in a thermal power plant, with photos or sketches of the main components

1.2.4 Reactors, boilers and exhaust purification systems

For generating the heat for the steam cycle, fuel is burned either in a gas-turbine, a boiler and a nuclear reactor. In nuclear reactors, a nuclear chain reaction of uranium is initiated, controlled, and sustained at a steady rate. To avoid a radioactive contamination of the environment, all components that come in contact with the radioactive material are shielded well. While manufacturing, service and maintenance of the radioactively contaminated components is done by specialized companies, the non-nuclear part of such power plants (steam cycle for power generation, including turbine and generator) are almost similar to coal and gas turbine power plants.

The boilers in power plants that burn coal, biomass or waste are quite similar to the oven in any normal house. However, the dimensions, temperatures and the amount of burned coal range in quite another dimension - with heights and diameters up to 100mxØ30m, temperatures between 800-1300°C and a daily fuel consumption of several tons. To avoid or at least minimize the pollution of the environment, the exhaust fumes are purified using several methods such as catalytic converters, filters or electrostatic precipitators. To also avoid the negative effects of carbon-dioxide to the atmosphere (greenhouse effect, global warming), currently the technology of "carbon capture and storage" (CCS) is exploited. This relatively new technology aims to compress the CO_2 generated in the power plant and store it in old oil- or gas-reservoirs where it cannot go into the atmosphere.

1.2.5 Components outside the power plant

Also the processes "before" and "after" the power plant contain sophisticated and quite valuable components. As normally these components result in relatively huge and complex steel structures as well, their needs for inspection, service and maintenance are usually similar to the components that can be found directly in power plants. Furthermore, these components have also been the first applications where climbing robots were successfully applied for inspection tasks.

For this reason, the following components are also considered in this work: Ships and pipelines for transporting the fuel to the power plant; and electrical cables that transmit the power to the end user.

1.3 Inspection of the core components

To guarantee a safe operation of the power plant, all components have to work reliable. For assuring this reliability, regular inspections have to be performed. As the economical losses that occur when a power plant is not producing energy are significant, the main goal of new technologies in this field should be to speed up the inspection time. As in comparable tasks (e.g. ship hull inspection) compact climbing robots have been already used successfully to substitute the expensive and time-consuming process of scaffolding; the idea to use this technology also in power plant inspection seemed straightforward.

1.3.1 Need for regular inspections

Since the first introduction of steam engines for power generation in the late 19th century, boiler explosions, generator crashes and other incidents in power plants have been a dangerous risk – always leading to severe financial losses and sometimes even significant demolition of the environment.

The biggest incident in the history of power plants was the one in Tschernobyl (Ukraine, 1986) that caused several hundreds of deaths and contaminated huge parts of the country until now. But also a relatively small accident in the "conventional" (= non-nuclear) part of a power plant, like the generator crash in Leibstadt (Switzerland, 2007) causes severe financial losses (approximately CHF 200 millions in this case) – and also deteriorates the public image of the nuclear power plants, as such accidents normally get exaggerated to a potential catastrophe by television and newspapers.

Accidents in non-nuclear power plants normally do not hit the headlines of newspapers in the same way as incidents in nuclear power plants – but of course they also happen. Even if the negative impact on the environment is normally much lower, also such accident create significant financial losses due to expensive repairs and long downtime without energy production.

For this reason, regular inspections are not only mandatory for the highly security-relevant components in nuclear power plants, but also highly recommended for other types of power plants and the "conventional" part of nuclear power plants – as the money invested in these inspections is normally much lower than the money that can be lost in an accident.

1.3.2 Economical impact

As power plant operators are aware of these risks, the service of inspection, maintenance and reconditioning has become an important business. For ALSTOM and other companies who are building power plants and components, the earnings generated by these tasks have become even more important than by selling new generators, turbines and boilers. However, the competition in this business is quite hard – not only with the original equipment manufacturers competing against each other but also some specialized companies penetrating into this market. For this reason, it is essential, to provide better technologies than the competitors for getting these highly valuable service contracts.

Regarding the economic point of view of a power plant inspection, one of the most important criteria is the overall inspection time. This criterion is mainly motivated by the fact that the downtime (no energy production) of a power plant causes losses in the range of approximately one million of Euro per day for a medium-size power plant. For this reason, the time consuming step of disassembling heavy parts should be avoided where it is possible. Also the risk of damaging these parts during the disassembly and the re-assembly process is an important argument when it comes to the development of new technologies that do not need the disassembly any more.

Other important criteria for high-quality inspections are repeatability and precision – repeatability for being able to establish the "history" of damages and thus do better predictions how long the part can still be used, and of course precision to better localize what to repair and thus only replace small parts instead of the entire components. Both these criteria can also be fulfilled better with modern robotic technology than with human operators. Additionally, through making the job of inspection more technological than dirty and dangerous, it also becomes easier to attract qualified young workers – which currently is a severe problem in this kind of business (average age of experts in NDT-testing > 50 years).

1.3.3 Limitations of conventional inspection methods

The most important limitations that make a direct human access dirty, dangerous or even impossible are the following: Narrow access space, long vertical sections or environmental hazards.

1.3.3.1 Narrow access space

In most cases, there is simply not enough space for a human to pass the entrance holes that are necessary for accessing the area of interest. For this reason, such components can only be inspected in a "conventional" way if they are opened or disassembled – with the disadvantages already described in the previous section.

If the area to inspect in not too far away from the entrance hole and if the access geometry is not too complicated, bore-scopes can be used. This method is frequently applied for short tubes and for the visual inspection of turbine blades, but not very precise and limited to these very specific applications.

1.3.3.2 Long vertical sections

Another limitation for the access is long vertical sections in many components. To access the areas at large height, either a complex system of scaffolds has to be installed or the worker has to climb to the point. While installing a scaffold takes several days of work, climbing to the point using ropes, ladders or similar equipment is dangerous, unpleasant and/or even unsuitable if the workers are old and their physical fitness is not excellent any more.

1.3.3.3 Environmental hazards

In some environments, the access is also limited by environmental hazards. Examples for such hazards can be the radioactivity in the inner part of a nuclear power plant, hot temperature, extreme noise or toxic ash particles in the air. For this reason, it is not recommendable to let humans work there.

Another type of hazard can also be the presence of water within the entire component – as it is the case in condensers or in water pipes. To face this problem, the use of waterproof robots can normally be cheaper than releasing all the water; and less dangerous than accessing the part with a human diver.

1.3.4 Advantages of inspections with mobile robots

Using mobile robots for the inspection can normally solve the above mentioned challenges in a very elegant way:

If designed compact enough, robots are able to enter spaces that are too narrow for humans and have a geometry that is too complicated to access with borescopes. If the mobile robots are even able to climb or fly, they can reach large heights and thus avoid the risk that a worker could fall down and get injured. Also external hazards such as heat, radioactivity or toxic ashes are normally less problematic for a mobile robot than for a human worker.

1.3.5 Focus on compact climbing robots

In this work, mainly the combination of the first two of the above mentioned challenges is addressed: accessing narrow spaces with complex geometry and long vertical sections, which requires compact size, vertical mobility and the additional ability to deal with specific geometric obstacles.

1.3.5.1 Compact size

With "compact" in the context of this work, we stress on robots that allow for reaching areas that are impossible to reach by humans due to the size restriction.

1.3.5.2 Vertical mobility and obstacle-passing capability

If the robot has to be very compact because of the narrow size restriction, the vertical sections within the component normally become relatively big in relation to the robot size. For this reason, most environments in power plants imply the need for vertical mobility.

Additional to the ability to move on smooth vertical or overhanging sections, very often also some specific types of obstacles such as corners, steps or ridges have to be passed. This combination of both moving with any respect to gravity and passing specific types of obstacles is called "climbing with 3D-mobility". The difficulty and variety of obstacles that can be passed by the robot define its grade of mobility – which will be analyzed very detailed in Chapter 4.

1.3.5.3 Combination = Compact climbing robot

While the possibility to use flying robots has to be rejected in most applications because of the narrow size restrictions and the relatively low payload capability of miniature air vehicles, the only remaining possibility is to use climbing robots – with the additional challenge to realize them at a size that is much smaller than in previous work where the only goal was to reach large heights.

For clearly distinguishing from other works which use the term "climbing robot" also for rough-terrain-rovers (Octopus [89]), cranes with suction cups (SIRIUSc [90]) or swimming robots with magnets (PoseiBot [94]), a "compact climbing robot" in the context of this work is defined as follows:

Mobile vehicle that is significantly smaller than a human, always stays in physical contact to the solid parts of the environment (= not flying or swimming), and is able to move in all inclinations in respect to gravity (not only on inclined slopes or rough terrain obstacles). The additional ability to pass specific obstacles is called 3D-mobility, with the grade of this mobility defined by the difficulty and variety of these obstacles.

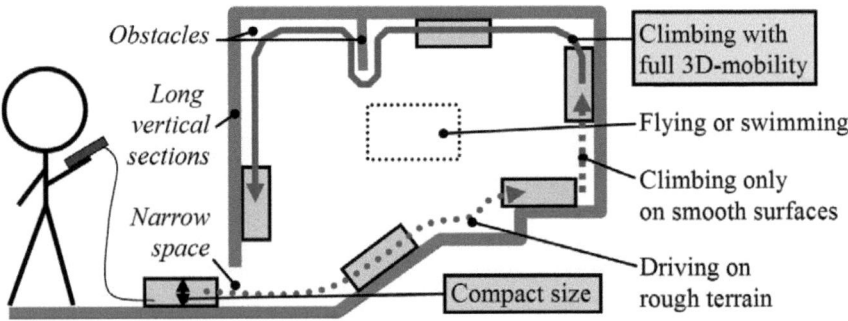

Fig. 1-5: How compact climbing robots are defined in the context of this work and how to distinguish climbing from other locomotion types

2 Requirements and functions

This chapter provides a brief overlook on the main challenges that are given both by the environment restrictions and the market requirements. Based on these requirements, an abstract functional analysis of typical compact climbing robots is provided, concluding with preliminary criteria and basic preferences that already focus on the basic principles that seem the most promising in the field of power plant inspection: Magnetic adhesion and wheeled or hybrid locomotion.

2.1 Environment restrictions

The environment restrictions are given mainly by narrow geometries that limit the maximum size of the robot, vertical sections that are often higher than this maximum size, and obstacles that cannot be passed with relatively simple vehicle structures. Sometimes, also the surface characteristics complicate the use of specific principles for adhesion or locomotion in the robot.

2.1.1 Overview on typical environments in power plants

As already stated in the first chapter, there is a high diversity concerning the components in power plants. However, some characteristics are quite similar in all of them. The main common characteristic is the fact that almost all of them are made out of ferromagnetic steel that allows using magnets for adhesion to the surface. Also the challenges coming from size restrictions, obstacles and surface-characteristics never reach the maximum difficulty in all 3 fields: Environments with very narrow size restrictions fortunately do not have extremely difficult obstacles (e.g. boiler tubes), while the environments with the most difficult obstacles (e.g. ridges in gas storage tanks) do not require very limiting size restrictions. For this reasons, accessing most of here described environments with compact climbing robots seemed challenging but still feasible when this work had started. Fig. 2-1 shows a selection of nine different applications in power plants. All of them have been analyzed in the context of this work; the four most interesting ones in more detail (a, g-i, highlighted with blue frames). A brief description of the main challenges within these applications is provided in the following text in this chapter; more detailed case-studies for the four main applications can be found in the case-studies at the end of this thesis (Chapter 5).

14   2.1. Environment restrictions

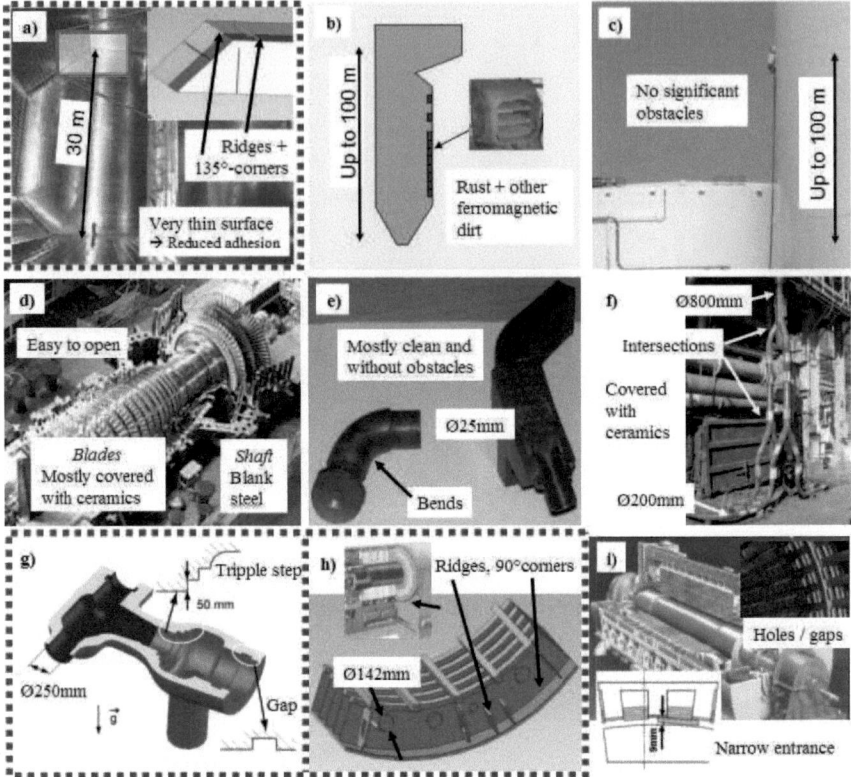

Fig. 2-1: Overview on typical applications and their main challenges – size restrictions, obstacles and special surface characteristics
(a) Gas storage tank in ships, (b) Coal-fired boiler, (c) Oil storage tank (picture taken from Jireh Industries [31]), (d) Steam turbine with blades, (e) Boiler tube, (f) Fuel pipe with intersections (g)Steam chest, (h) Generator housing, (i) Generator air gap

2.1.1.1 Storage tanks and boilers (Fig. 2-1: a-c)

Storage tanks and boilers form the biggest components to inspect in power plants. Because of their large size, there are almost no restrictions for the robot size and mass except the wish to transport it easily by one human operator and thus keep the mass below 20kg. However, the difficulties regarding obstacles and hazardous surface characteristics are sometimes very challenging:

The inner surfaces of gas storage tanks in ships (Fig. 2-1-a) are sometimes made out of very thin sheet metal that does not allow for very high magnetic forces because of saturation problems and have sharp ridges that can be seen as one of the most difficult obstacles for wheeled climbing robots. More details about this project can be found in the first case study at the end of this work (chapter 5.1).

In boilers (Fig. 2-1-b), the geometry of obstacles is less problematic, but the main problem is the fact that the surface is covered with several mm of rust and dirt. This dirt can have two negative effects, if climbing robots with magnetic adhesion are used: The non-magnetic part of the dirt reduces the adhesion force, while the magnetic part can stick to the robot and cause damage.

The application within this group that is the most friendly regarding the locomotion of climbing robots are the permanently installed (not on ships) gas and oil storage tanks (Fig. 2-1-c), as they can be found both in refineries and close to power plants. On these storage tanks, the surface is normally neither dirty nor contains difficult geometrical obstacles. They can be accessed with relatively simple climbing robots that only need vertical mobility, but no mechanism for obstacles and that are not restricted in their size. For this reason, such "classical" climbing robots are already in industrial use for these applications since several years, such as the Tripod designed by Jireh-Industries [31] or the MRS 200 by ALSTOM Inspection Robotics [32; Fig. 2-2-a].

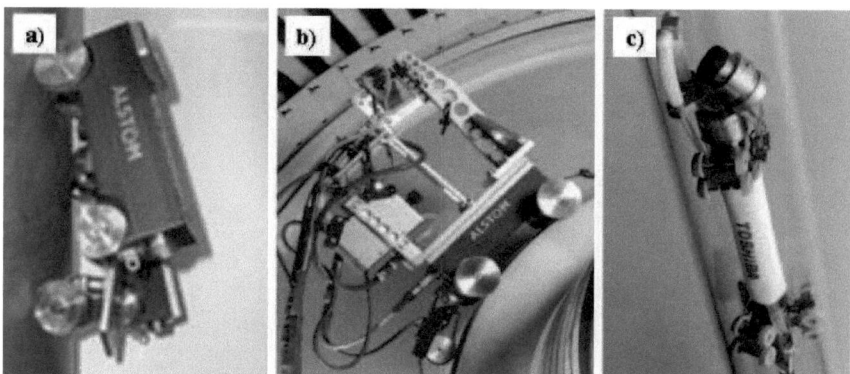

Fig. 2-2: Previous robots for power plant inspection
(a) MRS 200 [32], (b) MRS 100 [33] (both ALSTOM Inspection Robotics),
(c) Toshiba 1" tube crawler [57]

2.1.1.2 Turbines (Fig. 2-1: d)

Both gas and steam turbines consist of a rotor that is normally made out of ferromagnetic steel and several blades that are covered with a very thick layer of ceramics for heat protection and thus can be regarded as almost non-magnetic.

For inspecting these turbines, the "MRS 100" (Fig. 2-2-b and [32]) had been developed already before this work started. This robot consists of a relatively simple magnetic wheeled structure with only 1DOF for locomotion and an interface for several types of probe-holders and/or manipulator arms. It is placed on the shaft of an opened turbine where it is then moving on a circumferential path while its manipulator arm accesses the blades.

Climbing directly on the non-magnetic blades and/or accessing turbines with their housing closed can still be regarded as extremely difficult and almost unfeasible with current technology. It also does not seem very interesting from the economical point of view, as the turbines anyway get opened for each bigger inspection in order to replace some of the blades. A more detailed feasibility study for the idea to climb on the blades was once done by V. Hirschmann in his semester work at EPFL [103]. The concept to access the blades by using a relatively simple micro robot with 2D-climbing-mobility, an antenna pointing up to the blades and a folding mechanism for passing the narrow entrance holes was realized in a prototype at our team ([24], winner of the "Industrial Robot Innovation Award 2009", more details in chapter 5.4.3). Both concepts have not been followed until now.

Water and wind turbines have not been analyzed detailed in this work, but should also be mentioned briefly: While the inspection of water turbines should be very similar to gas and steam turbines, in wind turbines the main problem is to climb up the high towers – which sometimes are made out of non-magnetic material. An approach how to solve this challenge can be found in a recent study that was performed at London South Bank University [70].

2.1.1.3 Boiler tubes and pipes (Fig. 2-1: e-f)

In boiler tubes and pipes, the geometry is relatively simple compared to other environments: Circular shape that also allows for spreading against the other side to guarantee the adhesion in vertical sections (instead of using magnets); diameter changes mostly smooth; only sometimes turns and/or intersections. Also in this application field, several examples of robots with industrial relevance could be found already before starting this work, in all sizes starting from micro-crawlers for tubes with only 25mm (= 1") diameter ([57], Fig. 2-2-c) and reaching to quite sophisticated vehicle designs at bigger size that are able to negotiate complex-shaped intersections (e.g. the MORITZ by TUM, [55]). A very detailed overview on the state of the art in robots for in-pipe and tube inspection can be found in the pre-study for a project at CMU [100] – with focus on pipes with approximately 200mm; and in a study funded by ABB [101] – with focus on the smallest sizes of tubes. Our partners at EPFL currently also perform a study to improve the robots for small tubes in order to be more robust against dirt, water and other hazards [34].

2.1.1.4 Steam chests and generator housings (Fig. 2-1: g-h)

A very special case of pipe structure is last section before the hot steam enters the steam turbine. These so-called steam chests normally consist of geometry with several types of obstacles such as 90° corners, edges, tripple steps or holes. Furthermore, also the size of the robot is limited by pipes of only 250mm diameter, making it very challenging to realize a vehicle structure that can both pass the specified obstacles and is still small enough to fit through the narrowest pipes.

In generator housings, the geometry constraints are somehow similar, with the size restrictions even harder (entrance holes only Ø142mm) and some of the obstacles more difficult (ridges of 20mm thickness), but some of them also easier than in steam chests (only corners and edges instead of triple steps and holes).

While the inspection of stem chests is more important from the economical point of view (very expensive parts that are difficult to repair), the access of generator housings with compact climbing robot turned out to be the more challenging application regarding the vehicle design. Both applications are described more detailed in the case studies at the end of this work (chapter 5.2 and 5.3).

2.1.1.5 Generator air gap (Fig. 2-1: i)

In generator air gaps, the obstacles are less difficult than in most previously described applications – only gaps/holes that can be passed with a row of several wheels or a caterpillar. However, if the rotor needs to stay installed during the inspection, the entrance geometry becomes the narrowest of all applications – with only 9mm of total robot height allowed. Also the wish to not only perform paths in axial but also in circumferential direction is quite challenging regarding the vehicle design, as in this case the robot has to overcome gaps both in axial and in circumferential direction.

Given the high value of the generator and its central importance for the reliability of the entire power plant, the economical relevance of this application is the most important of all. Several robots already existed before this work started, but none of them could be applied in all types of generators. The limitations were either the very narrow entrance gaps in some ALSTOM generators or internal obstacles (zone rings) in GE generators. The development of a new robot for this application – with the goal to provide a universal solution for all types – will be described in the last case study within this work (chapter 5.4.1).

2.1.2 Size restrictions and height of vertical sections

As already shown in the previous section, the maximum allowed robot size is restricted in most applications: 9mm height in generator air gaps, Ø25mm in tubes, Ø142mm in generator housings and Ø250mm diameter steam chests.

Compared to these small allowed robot sizes, the height of vertical sections is always significantly bigger, which leads to another central requirement for a robot that is used for the inspection of power plant components – vertical mobility, which can be achieved either with flying or climbing. As the allowed space normally does not allow for flying, the main focus in this work is on climbing robots.

2.1.3 Typical obstacles

Beside being very small and having the ability to climb, an inspection robot for power plants also needs the ability to pass specific types of obstacles that normally differ a lot between the environments: Gaps/holes in generators, ridges in generator housings and gas storage tanks, triple steps in steam chests, tube intersections in fuel pipes.

It turned out soon, that the difficulty of an obstacle has to be regarded completely different when it is faced by a climbing robot instead of a rough terrain robot. Before this work, a comparison of obstacle-passing mechanisms had only been done in the field of rough terrain robots (e.g. the PHD of Thomas Thüer [106]) while in the field of climbing robots there was neither a significant number of mechanisms nor a detailed comparison yet.

Such a detailed classification and comparison of obstacles, mechanisms and vehicle concepts can be found in chapter 4 and can be seen as the main contribution of this work.

2.1.4 Surface characteristics and other hazards

Other challenges that directly result from the environment normally come from difficult surface characteristics that reduce the adhesion force. Examples are the rust cover in boilers, painted surfaces in some types of generators and thin sheet metal with saturation problems in gas storage tanks.

If the robot is not supposed to use magnetic adhesion but other principles, also other surface characteristics can become problematic, e.g. porous surfaces for robots that use pneumatic adhesion. These limitations are discussed more detailed in the description of these adhesion principles (Chapter 3.2).

Regarding other hazards, the presence of water or oil is the most critical – not only for preventing it from entering into the robot, but also because it significantly reduces the friction coefficient between robot and surface.

2.2 Market requirements and additional needs

Not only the restrictions given by the environment are crucial for the design of an inspection robot, also the market requirements form important constraints. First of all, a payload has to be carried to successfully fulfill the inspection task. Furthermore, the robot has to be simple to use and to repair, and be robust enough for not getting damaged when used in harsh environment and by operators with few experience. Additional needs are universality and modularity – to access a high number of applications with relatively few robots or modular components within these robots.

2.2.1 Payload

Even if just moving around in the complex geometries of these environments is already a difficult challenge – for successfully completing an inspection task, moving at high mobility is not enough yet, as also additional sensors and electronics have to be implemented to guarantee full functionality and safe operation.

2.2.1.1 NDT-sensors

For finding and analyzing the defects in the part to inspect, sensors for non-destructive testing are required. While many superficial defects (e.g. burnt windings on a generator stator) can already be detected with pure visual inspection using a simple camera (~5-20g), defects below the surface need to be analyzed with more sophisticated methods. The principles that are the most frequently used are eddy currents and ultrasounds, both requiring devices with approximately 10g to 100g (depending on accuracy and thickness of the part). Note that non-destructive testing with ultrasonic sensors requires a coupling liquid – resulting in additional complexity for its supply and negative influences on the friction between the surface and the robot wheels/legs (= similar to driving on a wet road).

2.2.1.2 Localization and navigation

Also localizing the robot and finding its path in complex environments is a challenging task that can normally not be solved with only using the camera image. For this reason, additional sensors such as inclinometers, encoders on all wheels, distance sensors or even laser range finders are installed on some robots. While the smaller sensor types are normally in the range of only 1-10g, the Hokuyo laser range-finder on the MagneBike already weighs approximately 300g including the mechanics for its actuation [7] – which is already close to the maximum payload capability of this robot.

2.2.1.3 Power supply and communication

Another challenge related to the payload of an inspection robot is the question how to supply power and how to communicate between the robot and the human operator. Basically, there exist four options: Connecting all sensors and actuators with a separate cable (1); placing a microcontroller on the robot that steers everything onboard and is connected to the operator's computer with a bus-cable (e.g. LAN or CAN) (2); placing microcontroller + battery + an interface for wireless control on the robot (3); or including so much intelligence to the robot that it can operate fully autonomously (4).

The last two options (3 and 4) are normally not applicable in power plant components, as safety restrictions always require a human operator for every device; and wireless communication signals could disturb other components and would anyway be shielded by the steel structure of the component. Within the other two possibilities, the simple option with separate cables is normally applied in very small and simple robots with little payload capability (e.g. the robots for generator housings), while the more complex alternative with a micro-controllers is applied on bigger systems that sometimes anyway need on-board-computing in order to allow for a real-time–control of critical functions (e.g. the stability control in the MagneBike [7]).

Regarding the mass of such components, in the case of the industrial version of the MagneBike all electronic hardware components for controlling 5 motors and 5 sensors sum up to approximately 230g – which is only 7% in the case of this robot (3.5kg) but would be almost unfeasible to carry by smaller robots that were developed for narrow environments (e.g. the Micro-Tripod WpW [14] – 50g).

2.2.2 Safety, reliability and robustness

Also safety, reliability and robustness are very important quality criteria for such inspection robots – especially when it comes to industrialized prototypes. For this reason, simple mechanisms and vehicles structures, with few parts that can get damaged are normally preferred to highly sophisticated but error-prone designs with many features.

2.2.3 Universality and modularity

As the number of different environments in power plant components is relatively large, realizing specialized robots for each application would be very expensive. For this reason, the ideal inspection robot should be as universal as possible and able to deal with a large number of different applications.

Given the large diversity within the applications (9mm allowed height in generator air gaps, 100kg payload for wind-turbine-inspection), a fully universal robot for all of them looks however quite unrealistic. For this reason, a fleet of different robots will always be necessary for covering the entire range of applications.

For keeping the overall complexity in such a fleet at a reasonable level, a good strategy is to stress on a modular design during the entire development process – if not possible in the mechanical design, then at least in the field of electronics and control. A good example for modularity in the mechanical design is the MRS-units [32, 33] that are used both in a vehicle for turbine blade inspection and in a 3-wheeled robot for storage-tank-inspection (Fig. 2-1-a and b). In electrical design, modularity leads to the use of similar motor controllers in the entire fleet – totally independently for which robot.

2.3 Main functions of a compact climbing robot

As already stated in the previous section, the ability to move to all desired points in the component – without opening or disassembling it – is the most crucial. This challenge had not been solved satisfactory for most application when this work had started. For this reason, the basic function "climbing with high mobility" became the main research focus. The other main functions of an inspection robot are analyzed as well, but only with the focus on their influence on the robot's mobility and its size.

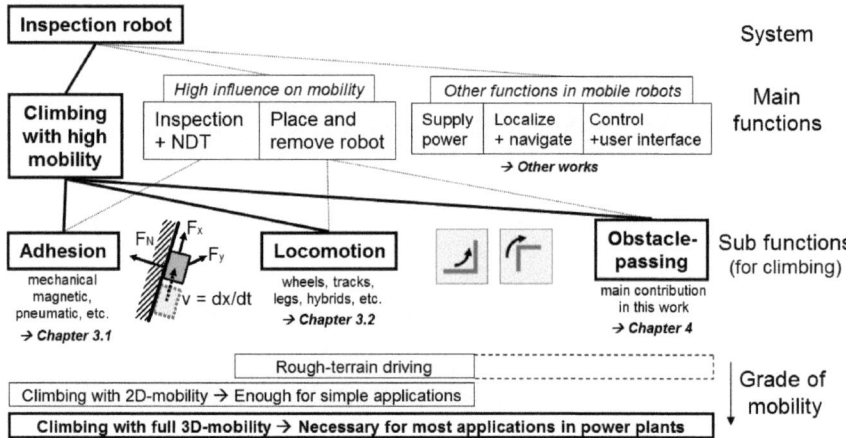

Fig. 2-3: Abstract functional analysis of an inspection robot, with the focus on how to climb with full 3D-mobility

2.3.1 Climbing with high mobility

As it can be seen in Fig. 2-3, three sub-functions are necessary for climbing with full 3D-mobility: Adhesion, locomotion and obstacle-passing.

2.3.1.1 Adhesion

For keeping the robot on the surface also in vertical or overhanging sections, the normal force (F_N) has to be increased. This function is called "adhesion". The most important adhesion principles in the field of compact climbing robots (mechanical, magnetic, pneumatic, others) and their influence on locomotion, mobility and size are described, analyzed and compared in chapter 3.1.

2.3.1.2 Locomotion

The basic sub-function for moving a device without external help is called "locomotion". This function can be found in all types of self-driven mobile devices, independently if they are able to climb or not. It is normally realized with legs, wheels or combinations, while some robots also use quite exotic principles (e.g. snake-like locomotion). A detailed description and classification of the numerous locomotion principles and their influence on mobility and complexity of the robot can be found in chapter 3.2.

2.3.1.3 Obstacle-passing

As already explained in 2.1.3, most environments contain difficult obstacles such as steps, ridges or sharp edges – which cannot be passed with simple vehicle structures. For passing them also with relatively simple robot structures, additional mechanisms are normally necessary. The detailed description, analysis and comparison of such mechanisms are addressed in chapter 4.

2.3.2 Other main functions in an inspection robot

Among the other main functions in a compact inspection robot, only two of them have significant influence on the robot's mobility - non-destructive testing (NDT) and placing/removing the robot.

2.3.2.1 Non-destructive testing (NDT)

As already stated in 2.2.2.1, the method of non-destructive testing can have a significant influence on the robot's mobility. One important influence is its mass and size (→ payload) that is very often not only the sensor alone, but also a small manipulator or probe-holder for moving it to the desired position. Additionally, some types of sensors also require a continuous movement of the robot (e.g. eddy current sensors) or use a coupling liquid that has a negative influence on the friction coefficient between robot and surface. More information about different sensors, their influence on the robot's mobility and the design of probe holders can be found in the Master thesis of Joaquín Catalá [104].

2.3.2.2 Placement and removal of the robot

Another function with high influence on the necessary robot mobility is the way it is placed in the environment to inspect. Very often, difficult obstacles in the entrance area can be avoided by just placing the robot after these obstacles – allowing for significantly smaller and simpler vehicles. Examples can be found both in the generator housing, the generator air gap and in steam turbines. More details about innovative solutions that facilitate the robot passing the obstacles at the entrance can be found in the chapter about these mechanisms (4.2.3.3) and in the corresponding case-studies. Note that also the idea of using two robots in mother-child-configuration can be seen as a form of "advanced placing method" for putting the child robot to its final destination.

2.3.2.3 Functions with low influence on the robot's mobility

The influence on the robot mobility that is given by the other main functions (power supply, control, user interface, localization and navigation) is relatively low and only interesting from the payload-point-of-view (see 2.2.1). More details of how to realize these functions in a compact climbing robot for power plant inspection can be found in the PHD-thesis of Fabien Tâche [105].

2.4 Main goals for the research

Summing up the most important requirements and their influence on the typical functions in a compact climbing robot, the main research goal for designing compact climbing robots for power plant inspection can be outlined as follows:

> High mobility – for dealing with many types of different obstacles,
>
> at low complexity – for realizing robust vehicles at very small size.

This central goal is very similar in most applications in power plant inspection, challenging from the technical point of view and had not been solved satisfactory for any of the analyzed applications before the project start. It can be seen as the recurrent theme that runs through all chapters of this thesis:

Chapter 3 (which provides a very general overlook on all types of climbing robots) finally shows which combinations of adhesion- and locomotion principles allow for the best compromises between high mobility and low complexity. The comparison of obstacle-passing mechanisms, vehicle concepts and prototypes in chapter 4 is as well based on these two core parameters. In the final chapter about the case studies, both abstract goals are then concretized based on the combination of obstacles and size restrictions that can be encountered in real power plant applications – showing both how the technical challenges have been solved and what economic benefit could be generated with innovative robots.

3 Classification of climbing robots

The goal of this chapter is to provide a comprehensive overlook on the most commonly used types of climbing robots – with a detailed focus on the main requirements in power plants: Very narrow size restrictions, mostly ferromagnetic, several types of obstacles and other types of hazards.

It is structured by the two main functions of a climbing robot – adhesion and locomotion – and points out to a rough classification and performance comparison that leads to the preliminary choice for the more detailed analysis in Chapter 4 – robots with passive adhesion, roll-legged locomotion and extra-mechanisms for passing specific obstacles.

3.1 Adhesion principles

As already described in 2.3.1.1 and Fig. 2-3, for climbing along vertical or even overhanging sections, the normal force (F_N) between robot and surface has to be increased to assure enough friction for holding the robot on spot and generating a traction force (F_T).

A comprehensive and detailed overlook on the most common adhesion principles was already done by D. Longo from Catania University and has been presented as a plenary talk on CLAWAR08 [98]. According to this classification, adhesion principles can be mainly distinguished among two criteria:

- Physical principle generating the adhesion force (e.g. magnetic, pneumatic)
- Energy need for generating the adhesion force (passive or active)

Also in this work, the most common adhesion principles are briefly described, analyzed and compared. However, a stronger emphasis is put on the specific requirements in power plant inspection – which normally necessitate very compact, robust and relatively simple solutions. What can also be observed in these environments is the finding that most structures are made out of ferromagnetic steel and/or consist of geometries that allow for mechanical adhesion. For this reason, mechanical and magnetic adhesion is described a little bit more detailed than the other principles.

3.1.1 Mechanical adhesion

As already explained in the introduction, many environments in power plants allow for increasing the normal force between robot and surface by just taking advantages of the specific geometry. This way of adhesion – called "mechanical adhesion" in most literature – is also the way humans and most animals climb; and the most frequent one in climbing robots with industrial relevance - especially in the group that spreads in tubes, pipes or gaps.

3.1.1.1 Inspiration from nature and mountaineering

For better understanding the different sub-principles within mechanical adhesion, a very intuitive way is to analyze how human climbers, monkeys or other animals hold themselves on vertical or even overhanging surfaces. Other inspiration can also be found in the belay devices that are used in mountaineering sports. In total, mainly four basic sub-principles exist, as it is represented in Fig. 3-1.

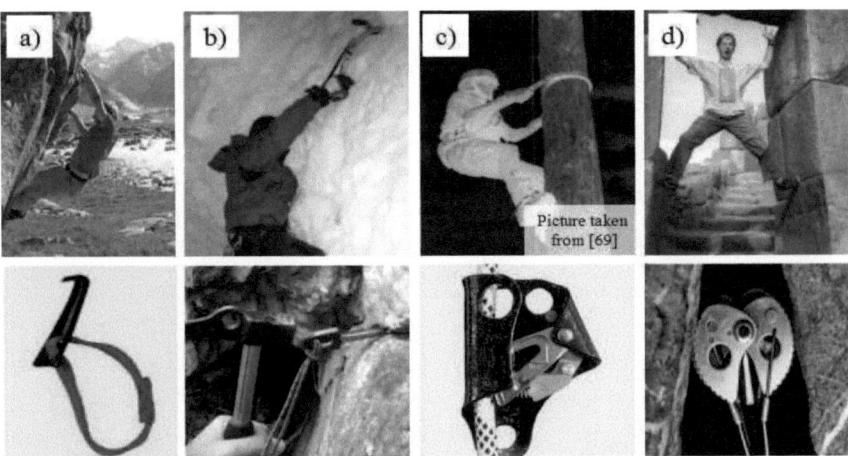

Fig. 3-1: Mechanical adhesion used by human climbers and in belay-devices for mountaineering (a) Form fit on specific surface features, (b) Penetration into soft material (e.g. ice), (c) Clamping on poles or ropes, (d) Spreading in a concave structure (gap or crack)

3.1.1.2 Basic sub-principles within mechanical adhesion

When analyzing these four sub-principles from a very abstract point of view, it can be seen that they can be grouped into "form fit" and "friction fit", and that very often combinations are used. Fig. 3-2 shows the simplified 2D-sketches of the four basic sub-principles plus two special cases that are used very often – both in mountaineering and also in climbing robots.

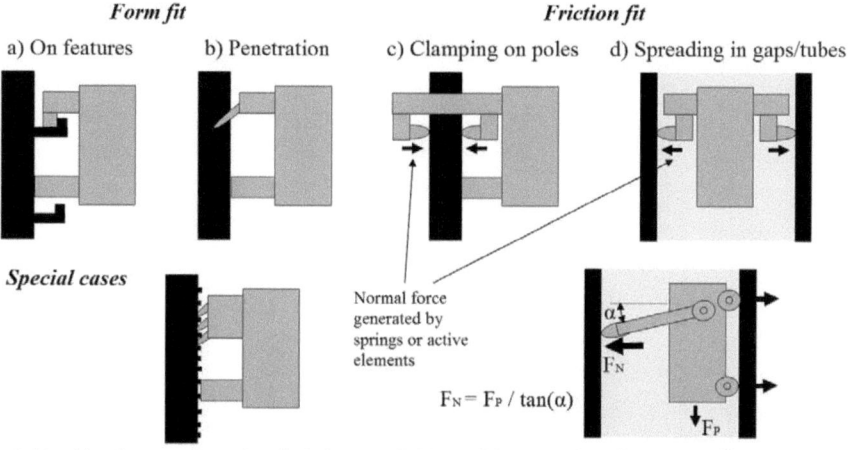

Fig. 3-2: Basic mechanical adhesion principles – Left: Mainly form fit (a) on specific surface features (b) through penetration into soft material (e) as a combination of both using spines – Right: Almost pure friction fit (c) on poles or (d) in gaps/tubes, with the normal force necessary either generated by springs, active elements or by (f) a pulling force that mainly comes from the gravity of the robot and/or its payload

The simplest approach to climb a vertical wall is to use features on the wall that allow for placing hands and/or feet. If this is not possible but the surface of the environment is relatively soft (e.g. ice or wood), the most common alternative is to penetrate into it using axes, crampons, nails or similar sharp tools. Also a combination of these two mechanisms is possible, using arrays of very small spines – penetrating into the surface at soft sections and fitting on small holes or ledges on harder sections. As all these principles mainly make use of a form fit to the surface, the payload is only limited by the rigidity of the connection parts between robot and surface – reaching very high values in the case of solid features such as the bars of a ladder. However, there are also some severe drawbacks:

Adhesion based on a form fit to features or on penetration is quite difficult to combine with robots that use wheeled locomotion and also limited to very specific surfaces (either with small ledges or similar features; or relatively soft). Furthermore, robots using this type of adhesion normally cannot move on the ceiling, as both spines and most types of surface features can only transmit high forces parallel to the surface but not perpendicular to it.

If wheeled locomotion is desired and/or the surface is smooth and without any features, the alternative within mechanical adhesion principles is to generate a friction fit by pushing against another surface that points in the opposite direction. This can be done either on convex surfaces (poles, rims, trees, ropes) or concave ones (cracks, gaps, tubes, pipes). For generating the normal force, normally springs, active elements or combinations of both are used. A very elegant way to increase the normal force (F_N) is to use knee levers, wedges, cam discs or similar mechanisms that transfer the pulling force generated by gravity (F_P) into a normal force towards the surface (cam disc: Fig. 3-1-d; knee lever: 3-2-f) – mostly even at a higher value than the pulling force itself. For also providing adhesion when the pulling force is small or not actuating in the desired direction, most implementations using this method also combine it with springs. Mechanical adhesion based on a friction fit is used in most climbing robots that move in tubes/pipes with vertical sections and/or that climb poles, trees or the towers of wind turbines. Implementations exist both with legs, arms and combinations. For generating the normal force, most robots just use springs or active elements; only few of them make use of a principle that transforms the pulling force generated by the gravity into an additional normal force.

3.1.1.3 Implementation in climbing robots

Fig. 3-3 again shows the four basic sub-principles within mechanical adhesion, this time in the context of typical prototype implementations in climbing robots. Again it is important to mention that only the mechanisms based on an almost pure friction fit can be combined with all types of locomotion principles, while the principles that are mainly based on a form fit can only be combined with swing-legged locomotion (legged or inchworm-type, see chapter 3.2.2) – resulting in a more complex control.

3. Classification of climbing robots

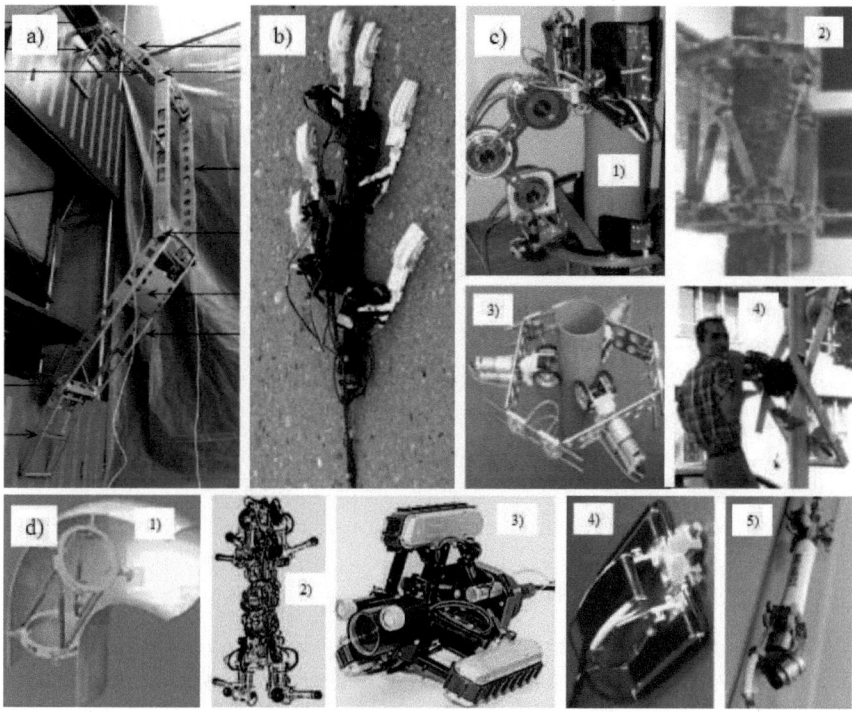

Fig. 3-3: Climbing robots that use mechanical adhesion principles:
(a) FireFighter [71], (b) SpinybotII [73], (c) Pole climbing robots
(1) 3D-climber [66], (2) SG-PCR [64], (3) Conceptual robot for windmill inspection [70], (4) Robot for climbing and cleaning lamp posts [69],
(d) Robots spreading in pipes, tubes or gaps (1) SG-PCR [64],
(2) MORITZ [55] (3) MicroTrac Vertical [56], (4) GE MAGIC [58],
(5) Toshiba 1" tube crawler [57]

The Fire-Fighter [71] profits of the handrails in balconies to access high buildings – somehow similar to a human climbing the steps of a ladder but at bigger size. Spines penetrating into the surface are used in the SpinyBot II [73] for not falling off the wall. With the spines realized at extremely small size and their shape and arrangement optimized for high holding forces, it is even able to climb vertical walls of concrete or very hard wood. However, overhanging sections or very hard and smooth surfaces such as glass cannot be climbed with this principle. Also wheeled implementations have not been reported – neither with robots using features on the wall nor with robots using spines.

In the field of pole climbing robots, the variation within the locomotion concepts is already much bigger. Implementations exist with most all types of locomotion principles – bipeds, wheels and combinations of them. The most frequently used configuration consists of 2 grippers that are united with an arm of several degrees of freedom (biped-configuration). A recent example is the 3D-climber [66] from Coimbra university, but also many other research teams have developed robots of this type (e.g. the InspiRat from Ilmenau [65] or the Roma 1 by Universidad Carlos III [67]. As an alternative to connecting the grippers with a standard serial arm, some robots also use parallel kinematics such as the SG-PCR developed at ETSII Madrid [64]. Note that this robot can also be adapted to inner tube inspection (Fig. 3-3-d-1)). Recent projects at London South Bank University (Robot for windmill-inspection [70]) and Teheran University (Robot for cleaning lamp posts [69]) have also realized wheeled robots for climbing poles, mainly driven by the goal to decrease the overall system complexity. The robot developed by the Iranian team even uses a mechanism that transforms the pulling force from the payload's gravity into an additional normal force – similar to the one described in Fig. 3-2-f and inspired by human pole climbing Fig. 3-1-c).

The biggest diversity of robots, and also the highest number of already industrialized ones can be found in the group that is spreading in concave environments – pipes, tubes or gaps. There, almost all types of locomotion principles can be found – multiple legs, bipeds, inchworms, wheels, tracks and hybrid solutions. Typical examples from the most important sub-groups are the in-pipe-version of the SG-PCR [64], the Moritz [55], the Microtrac Vertical [56], the GE Magic [58] or the Toshiba 1" tube crawler [57]. Several alternatives within the detailed design of the spreading mechanisms have been proposed as well – for dealing with highly different pipe diameters but still keep the complexity low. Also for passing sharp bends and intersections, several mechanisms have been proposed and implemented. A more detailed overview on in-pipe-inspection robots – mainly for pipe diameters around 0.1m to 1m – can be found in a study performed at Carnegie-Mellon-University [100]; more information about the smallest realized tube-inspection robots is provided in a similar study by ABB [101].

3.1.1.4 Industrial relevance and applicability to power plant inspection

Not only when counting the number of already realized robots, but also when looking at the smallest achieved sizes (down to only Ø10mm) – robots that mechanically spread in tubes or pipes are the type with the biggest industrial relevance. However, due to their need for always finding an opposing surface to spread against, they cannot be used for all applications – as they are either limited to a very specific tube diameter (e.g. the Toshiba-tube-crawler [57]) or result in a relatively big size if they are able to deal with large diameter changes or intersections (e.g. the MORITZ [55]).

Pole-climbing robots also have some industrial relevance and have already been proposed for inspecting wind turbines [70] or cleaning lamp posts [69]. Compared to in-pipe-inspection robots, their importance is however still smaller.

Climbing robots using large features, penetration or spines have been mainly built for exploratory purposes or military applications – and do not seem very promising for power plant inspection due to their high complexity, their large size and their limitation to walls that are only slightly steeper than vertical or contain enormous features.

3.1.2 Ferromagnetic adhesion

As mechanical adhesion suffers from the limitation to very specific geometries, most climbing robots for power plant inspection take advantage of the fact that most environments to inspect are made out of ferromagnetic steel – which allows for increasing the normal force between robot and surface by using a magnetic field that passes through the robot and the surface.

3.1.2.1 Electromagnets vs. permanent magnets

For providing such a magnetic field, basically two options are possible - electromagnets or permanent magnets.

Electromagnets have the advantage that they can be switched on and off at any time, making their use advantageous for robots that use swing-legged locomotion and need to easily release the feet for placing them in the new position. However, they require a constant energy supply for just keeping the adhesion (active adhesion principle) and are quite difficult to integrate into wheels or tracks.

For this reason, almost all recently developed climbing robots with ferromagnetic adhesion use permanent magnets – usually with an NdFeB-alloy. This alloy allows for very strong fields at very small size and mass, with values that are approximately 10 times higher than it can be achieved with hard-ferrite-alloys and almost comparable to electromagnets. Due to a significant price reduction in the recent years (mainly driven by Chinese suppliers) and the fact that these magnets can also be easily integrated into wheels and tracks, they almost completely substituted the electromagnets in the field of climbing robots. For changing between high and low adhesion force in the feet of swing-legged robots, several alternatives to electromagnets have been proposed and successfully implemented.

3.1.2.2 Position of the magnets

Regarding the position of the magnets within the robot, basically three possible configurations can be distinguished:

- In the feet of a robot using swing-legged locomotion
- In the chassis of a robot using roll-legged locomotion (see Fig. 3-9)
- Directly in the wheels, tracks or "special wheels" of such a robot

In robots with swing-legged-locomotion and magnetic feet, the feet normally are equipped with a mechanism for active force reduction – which allows for changing between the maximum value for the adhesion force and a significantly reduced value. This variability in the magnetic force is necessary for exactly defining which foot gets released before moving it to the next position and for reducing the actuator torque in the joints. For realizing such a foot with variable force, basically three possibilities are possible: Electromagnets, permanent magnets with lift mechanisms and special assemblies with a moving magnet. A detailed description and comparison can be found in the chapter about mechanisms for active force reduction in chapter 4.2.1.1. Examples for the implementation in robots can be found in the Robinspec [51] (electromagnets), our concept for helicopter-docks [23] (lift mechanism) or the magnetic switchable device developed by Aussie Kids Toys [52] (special assembly). Passive mechanisms for levering out permanent-magnetic feet without using an extra actuator have also been implemented in robots, but to our best knowledge did not allow for reliable mobility on difficult obstacles yet. For example, the MIT robot for bridge-inspection [50] sometimes falls off when it applies the gait form that is necessary for passing difficult obstacles such as steps, gaps or ridges.

In robots with roll-legged locomotion, basically two possibilities exist: Placing the magnets in the robot chassis; or directly in the wheels, tracks or special wheels.

The first magnetic robots with roll-legged locomotion mainly used wheels and magnets in the chassis, as in the time they have been developed the strong NdFeB-magnets in ring or plate shape (necessary for magnetic wheels) have not been easily available. One example is the Venom-Robot [30] that uses the magnet of an old hard-disc for its adhesion. Newer examples are the welding and inspection-robots developed at London South Bank University [29] or the newest version of the MRS 100 for turbine inspection [33]. Compared to robots on magnetic wheels, wheeled robots with magnets in their chassis bring the main advantages of not accumulating ferromagnetic dirt particles on the wheels and of not creating a magnetic forces to the side – benefits that make this technology useful in environments that are either very rusty or require to move the robot parallel to a ferromagnetic wall on the side (e.g. the blade ring of a turbine, see Fig. 3-4-b).

However, in more complex shaped environments – with the curvature radius of the topology being approximately the robot size or even smaller, magnetic wheeled robots seem to be clearly advantageous, as their adhesion force is less dependent of the environment geometry (see Fig. 3-4-c).

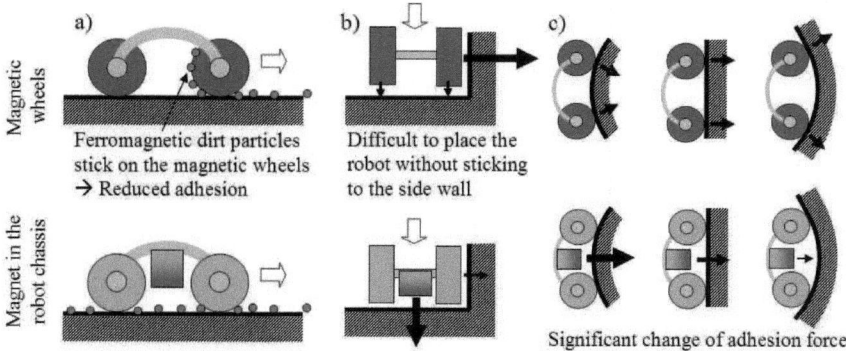

Fig. 3-4: Drawbacks (a, b) and advantages (c) of magnetic wheeled robots towards robots with magnets in the robot chassis: (a) Environment with ferromagnetic dirt particles, (b) Placement close to a side wall (c) Change of adhesion force in environments with curvatures

Also robots on magnetic tracks/caterpillars or special wheels (whegs) have been realized. More details about these concepts and their advantages and disadvantages towards wheels will be explained in the section about roll-legged locomotion principles and their implementation in compact climbing robots (3.2.1).

3.1.2.3 Basic mechanical structure of magnetic wheels

For realizing a magnetic wheel, the simplest way is to just fix a ring-magnet on a rotating shaft (see Fig. 3-5-a). By using good NdFeB alloys for the magnet, remarkable forces can already be achieved with relatively small wheels. For example, an Ø6mm ring magnet can already hold approximately 500g (or 5N) which corresponds to more than 100 times its own mass.

For conducting the magnetic flux on a more direct path into the surface and thus increase the magnetic adhesion force, most wheel designs additionally use two steel rims on each side of the wheel (see Fig. 3-5-b). Tests showed that the best results can be achieved with rims that are approximately as thick as the magnet – leading to a force increase of approximately 3-4 times. Another advantage of the design with rims is the increased robustness of the wheel – as massive steel is significantly more resistant against shocks than the sintered NdFeB-magnet-alloy.

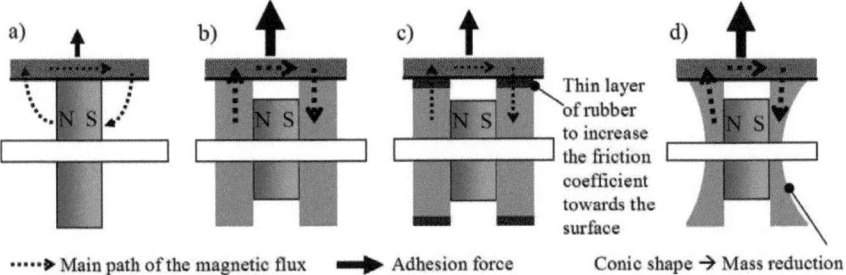

Fig. 3-5: Design and variation parameters for magnetic wheels (a) Most simple design using just a ring magnet, (b) Force increase by using two steel rims, (c) Design with a rubber tire for increasing the friction coefficient towards the surface, (d) Lightweight design with conic shaped rims

For increasing the friction coefficient between wheel and surface, magnetic wheels are sometimes covered with a thin layer of rubber (see Fig. 3-5-c). Even if the small gap in the magnetic flux that is produced by this "tire" decreases the adhesion force a little bit, higher traction forces can be achieved thanks to the significantly increased friction coefficient: In the case of the wheels for the MagneBike ([7], Ø60mm with 0.3mm rubber tires), the friction coefficient can get increased from $\mu \approx 0.3$ (blank steel rim) to $\mu \approx 0.8$ (with rubber tire), while the adhesion force is only reduced to approximately 70% of the value without tire. Another advantage of using rubber tires is the fact that they do not scratch the environment's surface. The main disadvantages are the relatively low lifetime of some rubbers on rusty surfaces; and the fact that the friction coefficient only gets significantly increased on dry surfaces (on wet surfaces, it also remains around $\mu \approx 0.3$). For this reasons, magnetic wheels that are mainly designed for wet or rusty surfaces normally use steel rims with a knurled (DE: "gerändelt") pattern on it – which achieve a friction coefficient of approximately $\mu \approx 0.5$ both on dry and wet surfaces but scratch the surface a little bit.

For decreasing the mass of the wheel without significantly reducing its adhesion force, a common approach is to use a conic shape for the rims (see Fig. 3-5-d). Tests with such wheels in the context of the gas-tank project (chapter 5.1 and [1]) showed that the mass reduction with this design can be up to factor 2, while the adhesion force only gets decreased less than 5%.

Most details of how to design and optimize magnetic wheels - with the example of the MagneBike wheels (Ø60mm) can be found in the PHD-thesis of Fabien Tâche [105] and in the JFR-paper about the entire project [7]. The values for other wheel sizes can be found in the corresponding design papers about magnetic wheeled robots that have been developed in the context of this work.

3.1.2.4 Industrial relevance of climbing robots with magnetic adhesion

The relevance of robots that use magnetic adhesion is not yet the same as with mechanical adhesion, but its significance is increasing:

Since many decades, robots on magnetic wheels are used for inspecting storage tanks, ship hulls or other environments with almost no size restrictions (e.g. the Tripod by Jireh Industries [31]). Also for generator air gap inspection, many examples with more than 10 years of history can be found – most of them running on caterpillars (e.g. the DIRIS Flex [36]).

More recent examples in this group are the MRS series by ALSTOM Inspection Robotics [32, 33] and the robots developed by our team – in collaboration with ALSTOM and BlueBotics. Out of these robots, already two of them have been industrialized successfully – the MagneBike [7] and the child-robot for the gas-tank-scenario [2]. Also other recently developed prototypes from our team are planned for industrialization. More details about these projects and prototypes can be found in the last chapter about the case studies (5).

3.1.3 Alternative adhesion principles

If the environment to inspect is neither made out of ferromagnetic steel nor contains a geometry that allows for mechanical adhesion, other principles have to be used for keeping the robot on the surface. The most common of these alternative principles are pneumatic, electrostatic and chemical adhesion.

To our best knowledge, all these principles show significant drawbacks towards magnetic or mechanical adhesion. For this reason, we have not found any existing robot in this group that looked promising for at least one of the analyzed applications (more details about the specifications, see chapter 5):

- Except swing-legged robots with pneumatic adhesion and sizes above 1m, almost no robot achieves enough mobility to pass inner and outer transitions in all possible inclinations.

- Almost no robots except some pneumatic ones based on vortex-technology are able to create enough adhesion on curved surfaces in pipes.

- Only few robots (e.g. the WaalBot [84]) have been realized small enough for passing the narrow entrance holes (Ø250mm or even smaller).

- Most passive principles (passive suction cups, gecko hairs, glue) have only been successfully tested on very smooth and clean surfaces; and even there usually achieved significantly lower adhesion forces than magnets of the same size

- The stronger and more robust principles (electrostatic and active vacuum suction) all need active actuation and thus need constant power supply.

For these reasons – and keeping in mind that almost all components in power plants are made out of ferromagnetic steel – the industrial relevance for power plant inspection is still very low. However, there has been done a lot of research into this direction and this work is still going on – usually with the goal to realize robots for search-and-rescue or military applications.

Perhaps, these technologies will once be improved in a way that they can be successfully applied in power plant inspection. For this reason, and to be complete, this subchapter provides a brief overlook on these alternative adhesion principles.

3.1.3.1 Pneumatic

Within these alternative principles, the most frequently used in mobile robotics is pneumatics (often also called vacuum suction).

3. Classification of climbing robots

For realizing a robot with swing-legged locomotion and pneumatic adhesion, the most intuitive way is to use passive suction cups – similar as the ones used on car-windows for fixing electronic devices, curtains or toys. An example for such a robot is the CLAUS (Fig. 3-6-a, [76]), developed at University of Osnabrück. For compensating the leakage of these suction cups at least on some types of surfaces, some robots use arrays of several passive suction cups that vibrate at low frequency. With this method, they can always use the pulling force of one suction cup for pressing the other one back to the wall. This technology, investigated mainly by researchers at University of Hamburg [77] is still under development and was not proven on rough surfaces such as brick-stone or concrete yet.

When pneumatic adhesion has to be applied on such relatively rough surfaces, normally a vacuum pump is installed on the robot that keeps the negative pressure in each cup at constant level if significant leakage occurs. Examples for such robots are the SpiderBotII from Catania University [74] or the Roma II by Universidad Carlos III (Fig. 3-6-b, [75]).

For combining wheeled locomotion with pneumatic adhesion, basically two possibilities exist: Sliding vacuum chambers (with a vacuum pump similar to the previously described robots) or vortex-based robots (using a fan in the middle that is shaped similar to a Pelton-turbine). Examples for sliding vacuum chambers are the first two prototypes in the Alicia-family (Fig. 3-6-c, [79]) or the CROMSCI [78]. Vortex-technology is applied in the City-Climber (Fig. 3-6-d, [82]) or the Alicia VTX [81]. Current examples showed that robots based on the vortex-technology can be realized much smaller, lighter and stronger than the ones using a sliding vacuum chamber.

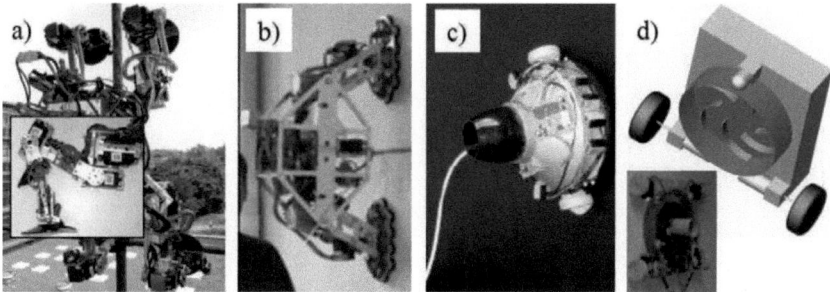

Fig. 3-6: Climbing robots with pneumatic adhesion (a) CLAUS - passive suction cups [76], (b) Roma II - active suction cups [75], (c) ALICIA 1 - sliding vacuum chamber [79] , (d) City Climber - vortex technology [82]

More details about pneumatic adhesion, with the history of their development, better comparisons of specific designs, recent trends and industrial applications can be found in the PHD thesis of D. Longo [108] and his overview paper at CLAWAR08 [98].

3.1.3.2 Electrostatic adhesion

Another method for increasing the normal force on non-magnetic surfaces is electrostatic adhesion. This principle was investigated more detailed at Stanford Research Institute (SRI) and implemented in the structure of a robot moving on caterpillars [86]. According to the researchers, this principle offers relatively high adhesion forces at low mass and reasonable energy consumption. As this technology is quite new and still under development, a detailed comparison towards other technologies cannot be provided yet. Perhaps it could become very promising in the future.

3.1.3.3 Dry adhesives and glue

Researchers at Stanford and CMU have also developed a special tape that is inspired by the feet of the gecko – an animal that is able to hold itself to the surface using a mixture of dry adhesion, electrostatic effects and van-der-Waals-forces. This tape has also been implemented into robots – the StickyBot [83] moving on six legs and the WaalBot [84] moving on whegs. Until now, these robots have only been successfully tested on clean glass surfaces – where also cheap passive suction cups would work well. However, the research teams promise that very soon this technology will also be applicable on more difficult types of surfaces.

Also the chemical reaction of glue can be used for the adhesion of climbing robots. Within the simplest type – adhesive tape placed on caterpillars or whegs – several robots have been realized, also in a previous project at our lab [88]. This principle suffers from the main limitation that the adhesion gets lost during time when the tape gets covered by dirt that is accumulated from the surface. For this reason, more reliable implementations of this principle also have an onboard cleaning system implemented on the robot – such as in the Gel-type Sticky Mobile Inspector [85].

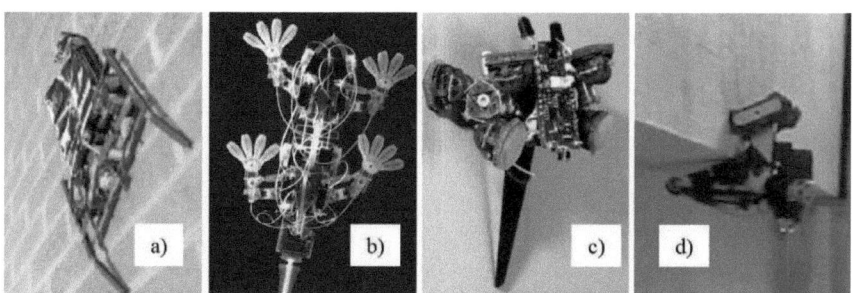

**Fig. 3-7: Climbing robots using other adhesion principles
(a) Electrostatic robot by SRI [86], (b) StickyBot [83], (c) WaalBot [84],
(d) Gel-type Sticky Mobile Inspector [85]**

3.1.3.4 Industrial relevance of these alternative adhesion principles

As already stated above, industrial applications have only been realized for robots with pneumatic adhesion – and their number is still small compared to magnetic or mechanical adhesion. To our best knowledge, the other principles have not been industrialized yet, mainly due to their low reliability, their limitations to swing-legged locomotion and specific surface characteristics and/or because the technology is not mature enough yet.

3.1.4 Active vs. passive principles

As already shown in the classification of D. Longo [98], another very important criterion for distinguishing the different adhesion principles is their need for energy consumption. While the active versions of each principle are normally much stronger than the passive ones, they suffer from the severe disadvantage that the robot would just fall down in case of a temporal power shutdown.

Mainly because of this reason, robots using permanent magnets (passive) have almost totally replaced their active alternative – electromagnets. However, in the field of other adhesion principles with less stronger force, the active versions are more common: Robots based on pneumatic adhesion normally use active suction cups, vacuum chambers or vortex-fans, as passive suction cups only work reliably on very clean glass surfaces due to leakage problems. Also in the field of electrostatic adhesion, only the active version works reliable – while artificial gecko hairs (normally using a mixture between Van-der-Waals-forces, dry adhesion and passive electrostatic forces) have only been proven on clean and smooth surfaces.

3.1.5 Comparison and performance evaluation

To sum up and to establish the link to the specific requirements in power plant inspection, the most relevant adhesion principles are compared against each other in Fig. 3-8. The adhesion principles are sorted by "passive" and "active", with the less relevant passive principles grouped. The performance criteria are mainly based on the typical requirements for power plant inspection. For better showing the main advantages and drawbacks of each principle, a simple color code is applied (shades of grey in BW-printing): Green (light-grey), if there are no severe limitations; orange (normal grey) if the limitations are not significant or irrelevant for a large number of applications in power plants; red (dark grey) if the limitation can cause significant problems in power plant inspection.

	Adhesion principle					
	Passive			Active		
	Mechanical (almost pure friction fit) 	Permanent magnets 	Others (spines, passive suction cups, gecko hairs, adhesive tape) 	Electro-magnets 	Active vacuum suction 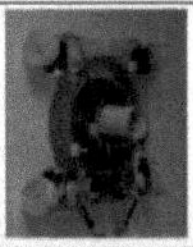	Electrostatic adhesion 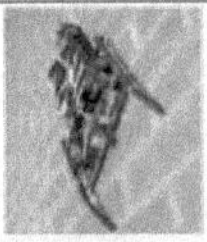
Power consumption	Nothing			Always		
Environment limitation	Specific geometry	Ferro-magnetic	Soft (spines) or very clean (all others)	Ferro-magnetic	Not porous	Nothing
Strength of adhesion force	Strong	Strong	Relatively low	Strong	Medium	Strong
Maturity of technology	Industrial use	Industrial use	Many research prototypes	Many research prototypes	Many research prototypes	Few existing prototypes

Fig. 3-8: Brief comparison of adhesion principles for climbing robots

As it can be seen quite easily, only mechanical adhesion in tubes or poles and permanent magnets are without any severe limitations regarding their applicability in power plants components while most other principles suffer from severe disadvantages. Within these alternative principles, all passive ones suffer from the main drawbacks that their adhesion force is relatively low compared to the others and that their specific requirements towards the surface are very limiting. These limitations are clean and smooth surfaces in the case of artificial gecko hairs, adhesive tape or passive suction cups, while robots based on spines can only climb on porous surfaces. For these reasons, they can be regarded as almost unsuitable for power plant inspection.

The active adhesion principles have the severe drawback that they always need power for just keeping the contact to the surface and thus can get problematic in case of a total power shutdown. Also the additional size of the adhesion zone can become a problem if the allowed size for the robot is very limited. However, if the environment to inspect is neither ferromagnetic nor allows for mechanical adhesion, these principles become interesting alternatives to the two main principles. Within the active principles, most research has been done in the field of vacuum suction, while robots that use adhesion based on electrostatic adhesion are still under development – with only a few existing research prototypes.

3.2 Locomotion

As already explained in 2.3.1.1, the basic sub-function for moving a robot without external help is called "locomotion". This function can be found in all types of self-driven mobile devices, independently if they are able to climb or not.

According to the well-known classification by M. Yim [97], terrestrial locomotion can be distinguished according to three basic criteria:

- Basic motion concept: **R**oll-legged versus **S**wing-legged
- Temporal characteristic of contact: **C**ontinuous versus **D**iscrete
- Type of contact: **Li**ttle footed versus **Bi**g footed

As for some configurations, a clear distinction in the type of contact (big or little-footed) cannot be done, 2 pairs of combinations can be grouped together - resulting in totally six combinations: RCL, RCB, RD, SDL, SDB and SC.

In this chapter, each type is described briefly – showing examples in the field of climbing robots and their performance according to the specifications in power plant environments. The first two sections deal with the basic types – wheels (RCL), caterpillars (RCB) and whegs (RD) in the field of roll-llegged locomotion; and multiple legs (SDL), bipeds (SDB) and snakes (SC) for swing-legged locomotion. Also hybrid solutions are described. The chapter concludes with a brief comparison of the most important types, pointing to their relevance in compact climbing robots for power plant inspection.

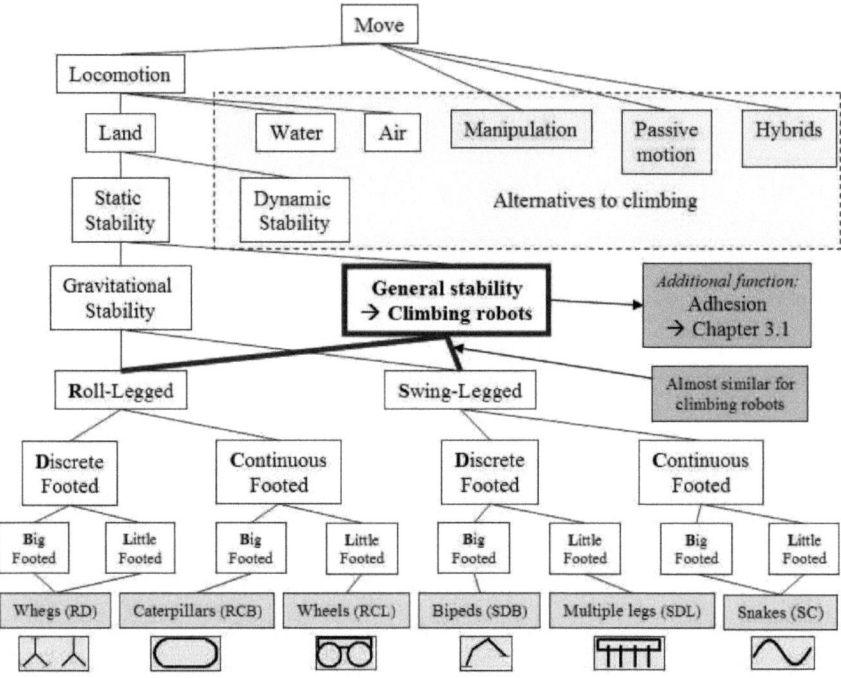

Fig. 3-9: Taxonomy of locomotion by M. Yim [97, white], extended by the main alternatives for locomotion (yellow), the necessary additions for climbing robots (blue) and the popular names plus images for the basic principles of quasi-static and ground-based locomotion (green)

3.2.1 Roll-legged locomotion

Within roll-legged locomotion, the most commonly used principle is to roll on wheels. Wheeled climbing robots allow for fast speed, simple control and low mass but also suffer from some disadvantages: Not all adhesion principles can be directly integrated into wheels, as they require a relatively big contact area towards the surface. Additionally, when a wheel comes into contact with two surfaces in an inner transition (corner) it can get stuck on the old surface. Also environments with small holes and gaps that have approximately the size of the wheels can get problematic if the wheels get stuck in these gaps.

For these reasons, some robots use caterpillars (often also called tracks) instead of wheels. While on straight surfaces caterpillars assure a large adhesion area that leads to relatively high forces, on convex curved surfaces the adhesion force gets significantly reduced and the caterpillar just gets peeled off – causing the robot to fall down. To our best knowledge, no research team has already found a reliable solution to solve this problem on all types of curvatures and outer transitions.

A relatively new approach to combine the advantages of wheels and tracks is locomotion on whegs (sometimes also called rimless wheels). In contrast to wheels and legs, in this locomotion principle the temporal characteristic of contact is not continuous but discrete. Usually, the pulling force generated by releasing one adhesion zone is applied for pushing the other one against the surface in order to improve its holding force. This additional effect makes this principle very advantageous for some specific adhesion principles such as passive suction cups, artificial gecko hairs or some types of glue. However, the motion of a robot on whegs is not as homogeneous as on tracks or wheels – with negative influences on odometry, the image of an onboard camera and NDT-measurements.

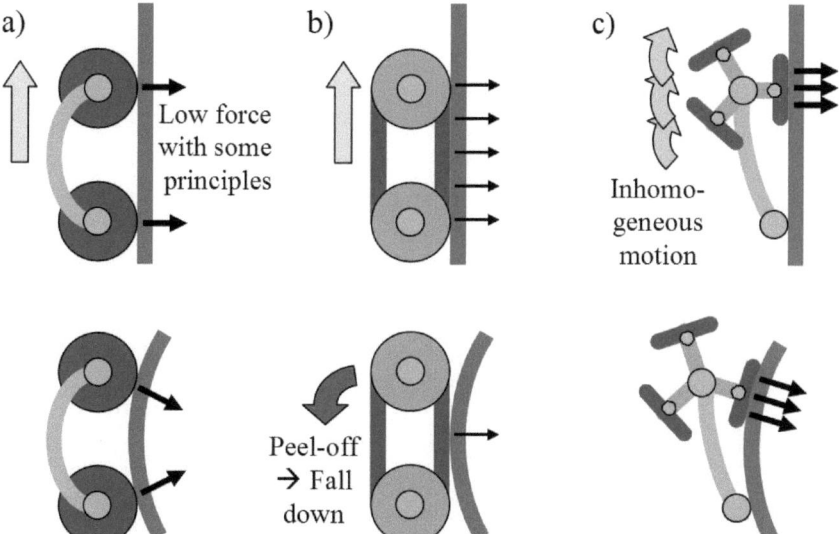

Fig. 3-10: Brief comparison of basic roll-legged locomotion principles when applied to climbing robots (a) Wheels, (b) Caterpillars, (c) Whegs

To sum up, all three roll-legged locomotion principles have advantages and disadvantages. Compared to swing-legged locomotion, their overall complexity is always very low, but their mobility in complex geometries is not as high. Locomotion on tracks or wheels can not be combined with all adhesion principles.

3.2.2 Swing-legged locomotion

Climbing robots that are based on swing-legged locomotion have been mainly developed for environments with a very complex-shaped topology and/or for applying adhesion principles that cannot be combined with wheels or tracks.

Implementations exist either with multiple legs and small feet (SDL) or in biped configuration with two big feet that cannot only transmit normal forces but also a torque towards the surface (SDB). While robots with multiple legs have been mainly realized to climb with relatively weak adhesion principles (e.g. passive suction cups, spines or gecko-hairs); biped robots are usually built for achieving a very high mobility on obstacles in complex-shaped environments. Most examples of this type have been realized with mechanical grippers for climbing on poles, but some examples also exist with active suction cups or electromagnets.

Note that within the robots based on the SDB-type, almost every research team uses different terms: Biped, inchworm, humanoid, brachiating robot, robot based on arms, caterpillar, double-gripper, and many more. In this work, we will mostly use the terms biped and inchworm. Biped, if the robot has the ability for plan-to-plan-transitions – and inchworm, if it can just move along one plane but is realized very flat (application: tubes and generator air gaps).

Some classifications (e.g. the one by D. Longo [98]) also distinguish between walking and frame walking – with the main difference that in frame walking some components within the robot move linear in respect to each other, while in "normal" walking all joints are rotational ones. As this detail within the mechanical design normally does not have a significant influence on the robot's performance, this differentiation is not done in this work.

To complete the group of swing-legged locomotion, also snake-like robots (SC) should be mentioned briefly. Even if mainly developed for rough terrain exploration, recent prototypes also succeeded in climbing vertical poles or in pipes by using the robot body for mechanical adhesion (e.g. the snake robot by SINTEF, ITC [60]). However, the control complexity of pure snake-like locomotion (without active wheels integrated into the segments) is still enormous and even worse when the snake robot has to climb. For this reason, it will be quite unlikely that they become relevant for power plant inspection within the next years.

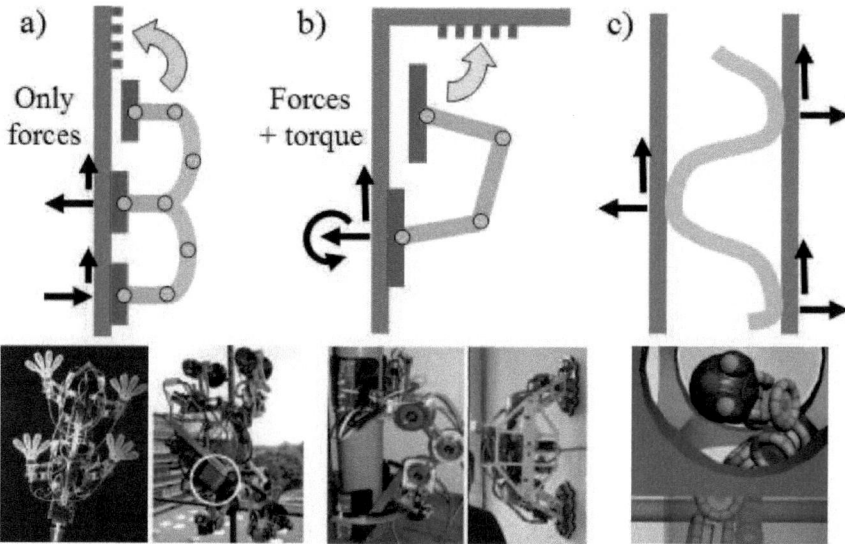

Fig. 3-11: Swing-legged locomotion principles and typical examples in climbing robots (a) Multiple legs for also applying relatively weak adhesion principles (StickyBot with gecko hairs [83], Claus with passive suction cups [76], (b) Biped robots for high mobility in transitions (3D-climber [66], Roma II [75]), (c) Snake robot spreading in a pipe [60]

What all swing-legged locomotion types have in common is their relatively high control complexity compared to roll-legged locomotion – normally resulting in a relatively big size and in slow speed. But what is gained for this high complexity is also an increased mobility, especially in the field of biped robots that look by far the most promising within the swing-legged group.

3.2.3 Hybrid locomotion

Also some robots have been developed that combine both swing-legged and roll-legged locomotion. Even if this approach looks quite complicated at first glance, it is a very interesting opportunity to profit from the advantages of both: high speed, high mobility and a complexity that is somewhere between roll-legged and swing-legged locomotion.

Examples for this type are the MrInspect [62], a pipe-inspection robot with several segments in snake-like-arrangement; the Alicia³ [80] which combines three sliding suction cups on an active frame; and the robot from our team which has been developed for the inspection of gas-storage tanks with thin ridges [1]. The generator air gap crawler with bi-directional mobility [18] uses wheeled locomotion in axial direction and biped locomotion with linear actuation in the structure for axial paths (often also called "inchworm-movement with frame-walking"). With this configuration, it achieves bi-directional mobility at only 8mm height and can deal with the relatively challenging topology of generator stators.

Fig. 3-12: Examples for hybrid locomotion in climbing robots
(a) MrInspect [62] (b) Alicia³ [80], (c) Robot for gas storage tanks [1], (d) Generator air gap crawler with bi-directional mobility [18]

3.2.4 Relative performance in mobility and complexity

As already pointed out in the introduction, the main goal addressed in this work is to realize robots with high mobility at relatively low complexity – for moving in complex-shaped environments but still staying at very small size. For this reason, a rough estimation of performance for each principle has been done according to these main criteria. This performance evaluation has not only been done in the field of climbing robots, but also for classic rough-terrain robots that do not use any adhesion principle – with the goal to better explaining the differences.

Fig. 3-14 shows this evaluation, using a very rough grading scale with 3 possibilities (low, medium, high) and highlighting the fields where the grade changes from classic rough-terrain-robots to climbing robots.

As it can easily be seen, some principles are quite different when regarding their performance in both groups – even if the final robots sometimes look quite similar; and at first glance the difference between classic rough-terrain locomotion and climbing does not become as evident.

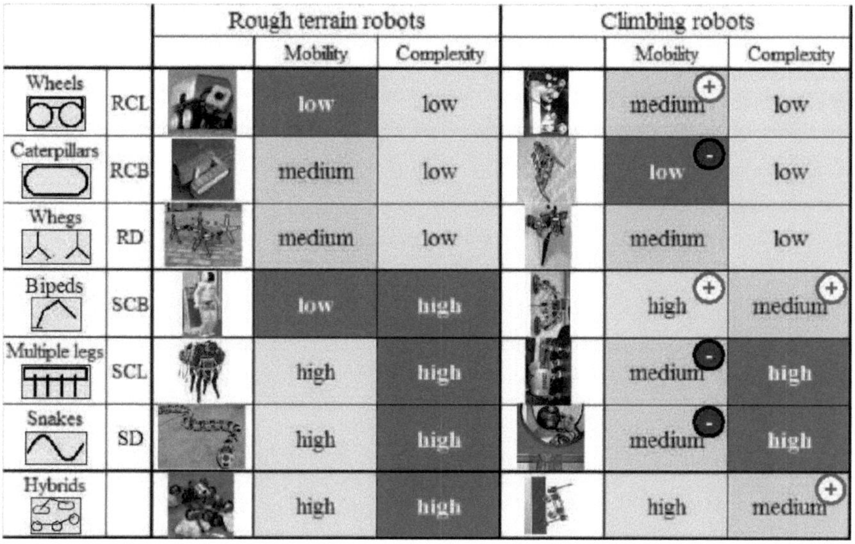

3. Classification of climbing robots

		Rough terrain robots			Climbing robots	
		Mobility	Complexity		Mobility	Complexity
Wheels	RCL	low	low		medium (+)	low
Caterpillars	RCB	medium	low		low (−)	low
Whegs	RD	medium	low		medium	low
Bipeds	SCB	low	high		high (+)	medium (+)
Multiple legs	SCL	high	high		medium (−)	high
Snakes	SD	high	high		medium (−)	high
Hybrids		high	high		high	medium (+)

Fig. 3-13: Rough performance evaluation of the most common locomotion types – both for classic rough terrain robots that do not use any adhesion principle and for climbing robots: Mobility on typical obstacles and complexity; + and - showing where there are significant differences between both groups; The pictures of most rough-terrain robots are taken from the slides of the course "Introduction to mobile robotics" [102]

Even if in some fields could of course still be changed as such a rough comparison can never be 100% objective, some tendencies can be observed:

3.2.4.1 Locomotion types with lower performance in climbing robots

Caterpillars, multiple legs and snakes are the locomotion types that mainly lose in their relative performance towards other principle when it comes to climbing in complex-shaped environments. However, the reasons are quite different:

While the big contact area of a caterpillar is a significant advantage when it comes to the negotiation of rough terrain obstacles (e.g. sand, mud, stones), this stiff structure becomes a severe drawback when climbing along convex curvatures with small radius or edges – as it normally leads to an unwanted peel-off for the entire robot (Fig. 3-10-b). As such curvatures can normally not be avoided when moving in power plant components, the mobility of locomotion based on caterpillars has to be regarded as quite low in comparison to the other principles.

In the domain of robots on multiple legs, the addition of an adhesion principle to each foot again increases the number of active DOF – which for this type of robot is already very high. Given the fact that the weight restrictions for climbing robots are normally much harder than in walking robots on the ground, most developers approach this conflict as follows: By reducing the number of motors to the absolute minimum and by applying very simple motion-control-algorithms that need little computational power, they still reach acceptable complexity – but very often are only able to move on smooth vertical surfaces without obstacles. In many cases, these robots are not even able to turn or to climb down, making them unsuitable for even the simplest applications in power plant inspection. Even if some of these robots at least manage to climb small steps, inner- or outer transitions have not been successfully shown yet.

Robotic snakes suffer from the same problem as robots on multiple legs – the extremely high number of active DOF. Additionally, most developers in this domain are driven by the idea of total mechanical modularity, which means to use always the same element for each snake segment and to improve the mobility only by intelligent control algorithms. Mainly caused by this restriction, the mobility of current snake robots did not reach the one of other robots with same size and complexity yet.

3.2.4.2 Locomotion types with increased relative performance

While the above mentioned types perform less good in climbing than in rough-terrain locomotion, bipeds, wheeled robots and hybrids are the locomotion types that increase their relative performance when it comes to climbing.

The most significant change can be seen in the case of biped locomotion: In this group, the implementations without an adhesion principle implemented (also often called humanoids) are highly unstable systems that are extremely difficult to control and not very reliable when moving on sand, stones or other different types of terrains. This relatively bad performance changes completely when a strong adhesion zone is put to the feet, as the improved stability does not only improve the mobility but also strongly decreases the control complexity. As these advantages are quite evident and have already been noticed by several other research teams, the number of realized prototypes and concepts in this group is quite high – covering almost all adhesion principles with subgroups: mechanical adhesion on specific wall-features (FireFigter [71]), on poles (Roma I [67], Inspi-Rat [65], 3D-Climber [66] and in tubes (SG-PCR [64]); magnetic adhesion based on electromagnets (Inchworm [49]), special assemblies [52] or pulling-magnet-devices [23]; pneumatic adhesion (Roma II [75]). Implementations with glue, artificial gecko hairs or electro adhesion do not exist yet to our best knowledge, but could also be realized once.

While the significantly increased performance in climbing robots is quite obvious in the field of biped locomotion – the relatively high mobility of wheeled climbing robots has not been fully identified by many researchers yet: With traction on two pairs of wheels, enough adhesion force in the wheels and a simple rubber cover to increase the friction towards the surface, a magnetic wheeled robot is able to pass both 90° inner and outer transitions (corners and edges) at any inclination in respect to gravity. This finding, plus a calculation theory and a simple test prototype to prove it in reality has just recently been presented by our team at CLAWAR08 [3]. With relatively simple extensions to the wheels or the vehicle structure, the newest generations of magnetic wheeled climbing robots even achieve this mobility without rubber on the wheels or with just traction on one wheel pair. More details about these extensions can be found in the chapter about mechanism for inner transitions (4.2.1). As with bipeds, adding adhesion to the wheels significantly increases the mobility of the robot, without causing severe negative effects regarding its control complexity.

Also the group of hybrid locomotion profits when it is applied to robots with climbing ability: With the mobility in the roll-legged part (mainly wheeled) already being at a "medium" level, only a few active joints in the structure are necessary to allow for passing even the worst combinations of obstacles – keeping the control complexity still at a reasonable level. For example, the first concept for gas tank inspection [1] only needs two additional linear actuators for passing ridges or other short zones with no possibility for adhesion – which is far below the values needed in pure legged robots. A detailed overview on the most common configurations for such hybrid wheeled/legged structures can be found in the chapter about active mechanisms for outer transitions (4.2.2.1).

3.2.4.3 Most promising principles for power plant inspection

For realizing very small and simple robots at high mobility, only the locomotion principles that look promising in both complexity and mobility have been analyzed in more detail: bipeds and hybrids, because they achieve high mobility at reasonable complexity; wheels and whegs, because they achieve very low complexity at still reasonable mobility.

As it will be shown later in the case studies (chapter 5), mobility for even the most difficult obstacles is not necessary in most power plant applications, while the restrictions regarding size and allowed complexity are usually quite hard in all environments. Additionally, a continuous movement of the NDT-sensor and a reasonable speed of the robot is very often desired. For this reason, the main focus in this work is set on wheeled robots – with the main goal to increase their mobility on specific obstacles to a level that is comparable with the other principles, but without sacrificing the advantage of their simplicity. As some of these obstacle-passing-mechanisms are very similar to whegs or hybrid locomotion, also these groups are analyzed in a more detail.

3.3 Overview and performance comparison

Similar to the classification by D. Longo [98], also this chapter concludes with a general overview on representative prototypes, which is depicted in a matrix – with the columns formed by the adhesion principles and the lines by the locomotion principles (see Fig. 3-15). In contrast to the above mentioned classification, this matrix also shows exemplary robot pictures instead of just a reference number, arranges the locomotion principles according to the well-established classification by M. Yim [97] and structures the adhesion principles based on their need of energy (active vs. passive) and their industrial relevance. Additionally, the robots that are already in industrial use are highlighted with a green background.

As it can be observed easily, the step from research prototypes to industrialization has only been realized in a few groups of climbing robots – mainly in the field of robots that use magnetic or mechanical adhesion in combination with wheels or tracks. In the relatively huge group of climbing robots with active vacuum suction, to our best knowledge only the CityClimber [82] has been commercialized - and only as a toy. Additionally, suction cups are also used in cleaning robots for high buildings such as the SIRIUSc [90] or the TITO-500 [91]. However, these robots do not climb actively, but are moved by a crane rolling on top of the building and use vacuum suction only for staying close to the wall.

Also within the groups formed by the other adhesion principles, no commercial application in the field of power plant inspection or similar tasks has been reported yet. Regarding the swing-legged locomotion principles, no robot has been industrialized either – not even in combination with magnetic adhesion.

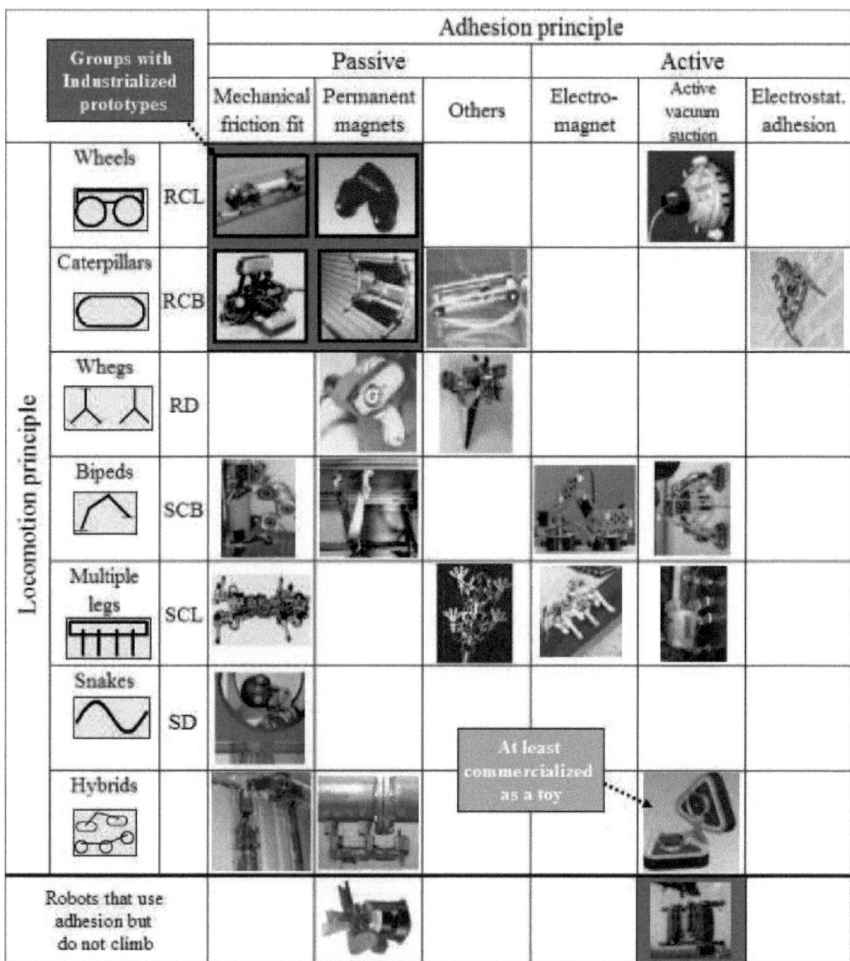

Fig. 3-14: Classification matrix for climbing robots – highlighting the groups where robots have already been industrialized
Reference robots on the pictures: Toshiba-Tube-crawler [57], MagneBike [7], Alicia 1 [79], MicroTrac Vertical [56], DIRIS Flex [36], TankBot [87], SRI robot [86], Rotating magnetic cam disc [15], WaalBot [84], Gel-Type Sticky mobile inspector [85], 3D-Climber [66], Robot for bridge inspection [50], Inchworm [49], Roma II [75], Moritz [55], StickyBot [83], RobInspec [51], SpiderBot [74], Snake by SINTEF ICT [60], MrInspect [62], PIR [48], City-Climber [82], Docking mechanism for micro-helicopters [23], SIRIUSc [90]

The main reasons, why only these few groups made it into industrialization have already been explained detailed in the previous sections, but are again summed up briefly:

- Only magnetic or mechanical adhesion allow for very strong forces at extremely small size and without a need for an extra power supply (passive principle).

- Roll-legged locomotion is simpler to control and more compact in the mechanical realization than swing-legged locomotion. The slightly lower mobility in comparison to biped robots is not critical in most applications. Additional mechanisms for specific obstacles even allow for reaching almost the same or even higher mobility in some environments.

- Also the high speed and the ability for continuous motion are significant advantages in wheeled and hybrid robots which are important for power plant inspection and similar tasks.

3.4 Conclusion and focus of the next chapters

In this chapter, the most important principles for adhesion and locomotion have been classified according to the two main functions for climbing (adhesion and locomotion); and compared according to typical requirements in power plant inspection. The method is based on the fusion of two previous works [97, 98] and enhances their scope by a detailed discussion according to the requirements in power plant inspection – bringing new aspects in the framework of a continuous enhancement of knowledge. At the end of each sub-chapter, comparison tables show in a comprehensive way, where research is already successfully transferred into industrial use, where there is still space for investigation and where one could get promising results in the future.

Considering the main indications given by these comparison tables, the last two chapters of this work mainly go into detail for the robot group that looks the most promising in the field of power plant inspection; but still offers challenges that are not completely solved yet: Magnetic wheeled robots – with the challenge to pass specific obstacles by using additional mechanisms without significantly increasing the overall complexity. In chapter 4, these obstacle-passing-mechanisms and robot concepts are described, classified and compared, while chapter 5 shows their application in exemplary industrial projects.

4 Obstacle-passing with compact climbing robots

In this chapter, the most typical obstacles and hazards for climbing robots are classified. This classification is followed by a description of the most common mechanisms that are used in climbing robots on magnetic wheels or other robot types that combine passive adhesion and roll-legged locomotion. The next section then provides an overview over entire vehicle concepts that combine the above mentioned mechanisms in order to achieve high mobility at reasonable size. The chapter concludes with a performance comparison of these vehicle concepts – where the performance of representative prototypes for each concept is evaluated against the most common requirements in power plant inspection. A more detailed technical description of these prototypes and the context of their development can then be found in the next chapter about the case studies.

4.1 Detailed classification of obstacles

Obstacles for climbing robots can be classified according to three basic groups – inner transitions (concave corners), outer transitions (convex edges), and curvatures. Also combinations exist very frequently – and are usually more difficult than each obstacle type alone. Additionally, specific hazards such as heat, rust, or water can occur as well.

4.1.1 Inner transitions (concave corners)

In inner transitions, the two main challenges are detaching from the old surface and assuring enough adhesion in the middle of the transition – when the robot chassis is the furthest away from the surface. Which challenge is the more difficult depends on the location of the adhesion zone: For robots with the adhesion zone is in the chassis, the reduced adhesion force at the middle of the transition is the most critical one, while for robots running on magnetic wheels the main problem is to detach the wheels from the old surface. The difficulty of an inner transition is mainly determined by 3 core parameters – inclination in respect to gravity (α), angle between the two surfaces (β) and friction coefficient between robot and surface (μ).

4.1.1.1 Detaching from the old surface

The main challenge for climbing robots on magnetic wheels is to detach from the old surface. To do this, the robot has to deal with an unwanted remaining adhesion force (F_{adhUW}) that is pulling back towards the old surface. As it can easily be observed in Fig. 4-1-b, the necessary traction force in this case is significantly higher than for just climbing on smooth surfaces (Fig. 4-1-a).

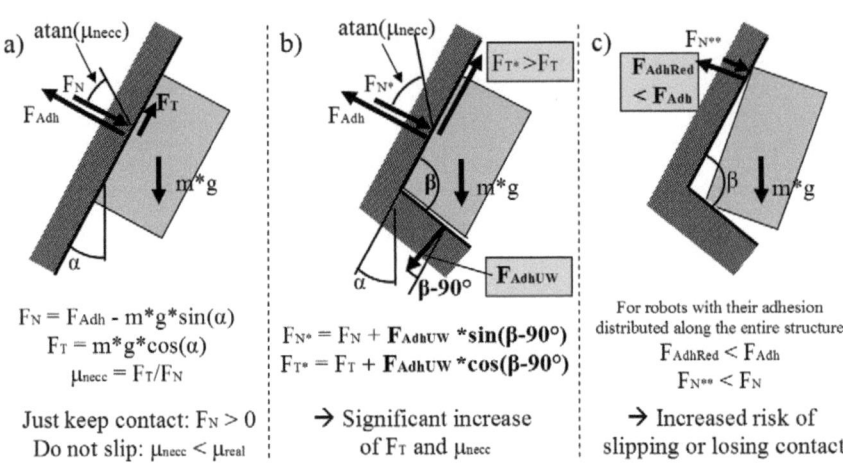

Fig. 4-1: Mechanical model of a climbing robot and its needs for adhesion force and friction coefficient towards the surface (a) Only climbing on smooth surfaces (= 2D-mobility), (b) Unwanted adhesion force when getting away from the old surface at the end of an inner plane-to-plane-transition, (c) Reduced adhesion force in the middle of the transition – critically for robots with their adhesion zone in the structure

More detailed calculations for magnetic wheeled robots can be found in our paper at CLAWAR08 [3]. This paper shows in calculations and real experiments, that a vehicle with four magnetic wheels, very strong gear-motors, traction on all wheels and a good rubber coating on the wheels to increase the friction coefficient (μ=0.6-0.8 instead of only 0.2-0.3) is able to pass such inner plane-to-plane-transitions in all inclinations. However, passing inner transitions with such a vehicle only works on dry and clean surfaces where such a high friction coefficient can always be assured. For not being limited to such environments, several mechanisms have been developed – mainly with the goal to decrease the unwanted adhesion force coming from the old surface (F_{AdhUW}). More details about these mechanisms are provided in chapter 4.2.1.

4. Obstacle-passing with compact climbing robots 55

4.1.1.2 Assuring enough adhesion during the transition

For robots with their adhesion zone distributed along the entire structure and not only around the wheels or whegs, another problem gets even more severe: The reduction of the adhesion force caused by the increased air gap between adhesion zone and surface significantly decreases the normal force – causing the robot to slip or even to lose contact (see Fig. 4-1-c). For this reason, most robots of this type (e.g. the Alicia VTX with vortex-based vacuum suction [81], the robot with structure magnets by London south-bank University [29], or the TankBot on adhesive caterpillars [87]) are only able to pass inner plane-to plane-transitions from ground to wall, but not from wall to ceiling - at least to our knowledge.

4.1.1.3 Core parameters for quantifying the difficulty

For better comparing the mobility of different robots against each other, the difficulty of inner transitions can be estimated based on the following parameters:

- Angle between the surfaces (β)

Definition: $\beta<90°$: acute angle, $\beta=90°$: right angle, $90°<\beta<180°$ obtuse angle. Generally, it can be observed that the bigger the angle β, the easier it is for a climbing robot to pass the inner transition – independently if the adhesion zone is distributed along the entire structure or only around the wheels.

- Inclination in respect to gravity (α)

As already mentioned before, most robots that have their adhesion zone distributed along the entire surface are only able to pass from ground to wall but not from wall to ceiling. On the other hand, for magnetic wheeled robots the worst case regarding the necessary friction coefficient is the transition from ground to wall. Some robots on whegs (e.g. the WaalBot [84]) however cannot move downwards in a controlled way, thus the "downward" transitions (ceiling to wall, wall to ground) cannot be performed. Giving these constraints, the inclination in respect to gravity is also a very important parameter for comparing the mobility of different robots in inner plane-to-plane-transitions.

- Necessary friction coefficient between robot and surface (μ_{necc})

Another important criterion regarding a robot's mobility in inner plane-to-plane-transitions is the minimum friction coefficient between robot and surface that is necessary for moving away from the old surface (μ_{necc}). Only if this value is significantly below the really measured friction coefficient (μ_{real}), the robot can pass the transition without any slip. If the values for μnecc and μreal are approximately equal, the robot can still pass, but the wheels normally slip for a short instant until they come into a position where the values for friction and/or distribution of adhesion are slightly more favorable than average. Important limit values for μnecc are at approximately 0.3 (= smooth steel wheel on dry surface or steel wheel with a knurled pattern on wet surface) and at 0.6-0.8 (= steel wheel with a relatively resistant rubber-cover on a dry surface, see Fig. 3-5-c). Note that the "real" friction coefficient (μreal) cannot be measured with high accuracy and strongly varies on many surface characteristics.

4.1.1.4 Sub-classes for the classification of inner transitions

For establishing a comprehensive classification that allows for an effective comparison of different robot concepts, the following sub-classes within inner plane-to-plane transitions have been derived – based on the most common requirements in power plant environments and on the main differences in existing robot prototypes: Angles 135° and 45° in all inclinations, 90° (the most common) split up according to two different inclinations (as many robots are only able to perform well on one of them). All these obstacles are meant for robots with rubber-covered wheels/tracks/whegs on dry steel surfaces (μnecc≈0.6); while the ability to also pass the specified obstacles without rubber cover (resistance against abrasive surfaces) and/or on wet surfaces (both: μnecc≈0.3) is mentioned as an extra additional property. Fig. 4-2 shows the symbols for the sub-classes within inner plane-to-plane-transitions that are used in the final comparison matrix for the performance evaluation of robot prototypes (chapter 4.3.3 and appendix).

Fig. 4-2: Sub-classes within inner transitions
(a) 135°, (b) 90° ground-wall, (c) 90° wall-ceiling, (d) 45°, (e) mobility also on wet and abrasive surfaces with μnecc ≈ 0.3
<u>Origin:</u> (a) Gas storage tanks, (b, c) Steam chest and generator housing, while in (c) some robots have problems, (d) Additional function offered by some robots, no link to applications yet, (e) Wish for more robustness and compatibility with NDT-methods using a fluid (mainly for steam chests)

4.1.2 Outer transitions (convex edges)

Also in outer transitions, several challenges have to be addressed. First of all, enough ground clearance has to be assured during all the transition, while the adhesion force should remain at an almost constant level – a challenge that until now has only been solved well with magnetic wheeled robots and robots rolling on whegs. Regarding the specific case of magnetic wheels, another important challenge is the significant reduction of the adhesion force on the edge caused by the saturation of the magnetic field – an effect that is limiting the payload in most climbing robots of this type. Similar to inner transitions, the difficulty within this obstacle group is also determined mainly by the angle and the inclination in respect to gravity. Additionally, also the curvature of the edge plays an important role when regarding the reduction of magnetic adhesion forces caused by saturation effects.

4.1.2.1 Ground clearance and other geometrical constraints

When passing an outer transition with a climbing robot, the most important challenge is to both keep enough ground clearance for not touching the ground during all the transition and to still keep the adhesion force high enough for not slipping. In robots with the adhesion zone integrated into the wheels, the only effect caused by this constraint is a slight reduction of the available space for payload, motors and gears (Fig. 4-3-a).

For robots with the adhesion zone integrated into the structure and/or robots with adhesive caterpillars, outer-plane-to-plane transitions are an obstacle that has – to our best knowledge – not been solved yet for all possible inclinations. On robots with the adhesion zone integrated in the structure (Fig. 4-3-b), a design with enough ground clearance during the transition would lead to a very large air gap in normal driving conditions – with not enough adhesion any more during normal climbing. On robots that use adhesive caterpillars (Fig. 4-3-c), enough ground clearance is not even possible at all, as a caterpillar is normally spanned tangential between the two wheels. For this reason, such a robot just peels off its caterpillar when rolling over the edge. Adding an elastic tail to such a robot (such as in the TankBot [87] and Fig. 4-26-b), can improve the mobility at least on the outer transition from a wall to the top ground, but to our best knowledge is not enough yet for also dealing successfully with the worst-case-transition from ceiling to wall (as it is drawn in Fig. 4-3-c).

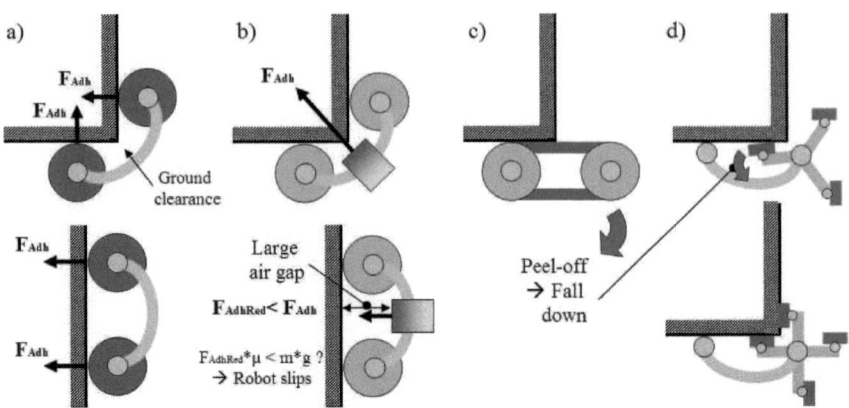

**Fig. 4-3: Challenges when passing outer plane-to-plane-transitions
(a) need for enough ground clearance in magnetic wheeled robots,
(b) reduced adhesion on robots with their adhesion zone in the structure,
(c) peel-off-effect in climbing robots rolling on caterpillars,
(d) extra geometrical constraints for designing a wheg**

Also with robots rolling on whegs a negative peeling effect can occur on outer plane-to-plane-transitions, especially if the wheg consists only of three lever arms and the joint between arm and adhesion zone is not flexible enough for this type of obstacle. In the WaalBot [84] this is very likely the case, as outer transitions are not discussed in any of the publications or videos on this robot; and the design seems to be only optimized for inner transitions. Wheg-designs that are better adapted to this type of obstacle are used in the gel-type Sticky Mobile Inspector [85] and in the hexagonal magnetic cam-disc [15]. Both design allow for both inner and outer plane-to-plane-transitions in all upward inclinations and are described more detailed in the next section. Also note that most current climbing robots on whegs are only able to move upwards; but not downwards (videos do not show it). For this reason, their mobility on outer plane-to-plane-transitions is of course only limited to inclinations where the robot is moving upwards.

4.1.2.2 Effects on the adhesion force in magnetic wheels

As already explained in the last paragraph, currently only robots on magnetic wheels or whegs are able to roll through outer transitions in all inclinations. For this reason, mainly this group is analyzed for establishing a more detailed subclassification according to the difficulty. Basically, two effects occur on a magnetic wheel when it is rolling over an edge during an outer transition:

4. Obstacle-passing with compact climbing robots

A significant reduction of the adhesion force in the middle of the edge (Fig. 4-4) and a force that is pushing the wheel away from the edge (Fig. 4-5).

The force reduction is mainly caused by saturation effects of the magnetic field in the zone of direct contact. Also the further increased air gap in the part of the field that already passes through the air plays a role in this force reduction.

In several test series with different magnetic wheels, the reduction of the magnetic adhesion force on edges has been measured – leading to the approximate values that are represented in Fig. 4-4: On rounded edges (with the curvature radius of the surface "$r_{surface}$" being in the range of approximately 0.1 times the wheel radius "r_{wheel}"), the force reduction results in approximately 50% of the value that is achieved on a plane surface. If the edge is extremely sharp (almost no curvature any more), the reduction can even go down to 20%. On thin double-edges (sometimes also called ridges or surface flips) or edges that are not well deburred, the adhesion force can even get reduced down to 10% or less.

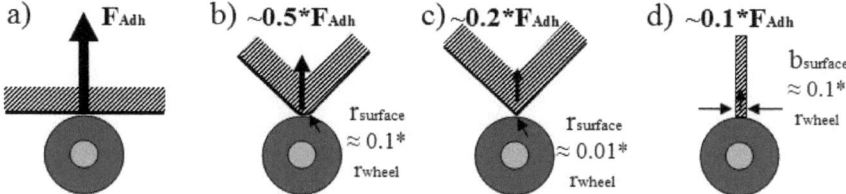

Fig. 4-4: Reduction of adhesion force of magnetic wheels rolling over convex edges or thin ridges an outer plane-to-plane-transition
(a) normal surface, (b) rounded edge, (c) sharp edge, (d) double edge

Taking into account these significant losses of adhesion force for the design of compact climbing robots, the following approaches have been followed to deal with this problem: Most robots such as the original MagneBike [7] or our first prototype for generator back housings [12] just use magnetic wheels that are extremely strong in comparison to the mass/gravity-force (factor 10 and more). In order to also deal with these high adhesion forces during inner plane-to-plane-transitions (see last section), very strong actuators and gears have to be chosen - normally resulting relatively large and heavy compared to the robot size and its payload capability. For completely avoiding the necessity to hold the robot with a wheel that is placed on an edge, many other robots such as the concept for GTT [1] and the MagneBike-PCAE [9] use more than two pairs of wheels so that the adhesion forces can always be distributed completely among wheels that are in full contact with the environment. Another approach to reduce the effect of force reduction on edges is to design special flexible wheels with less force reduction on edges (Semester work at ETH [47]). More details about these designs can be found in the chapter about mechanisms for outer transitions (4.2.2).

The other effect on magnetic wheels – the force pushing the wheel away from the edge – is however not resulting in a severe limitation for robot mobility and can even be exploited in a positive way. As it can be seen in Fig. 4-5, it is mainly generated by the asymmetric distribution of the adhesion force when the robot is on the edge. As its quantitative value is in the range of approximately only 10% of the adhesion force on a plain surface, it is normally no problem to overcome it on the beginning of the edge (Fig. 4-5-b). When leaving the edge, it is even acting in the direction that helps the robot to move forward (Fig. 4-5-d). Thanks to this effect, very short and lightweight vehicles such as the Micro-Tripod WpW [14] can even pass outer plane-to-plane-transitions without traction on the back wheels. A more detailed analysis of this effect can be found in the chapter about the technical details of the Micro-Tripod WpW (5.3.3.3).

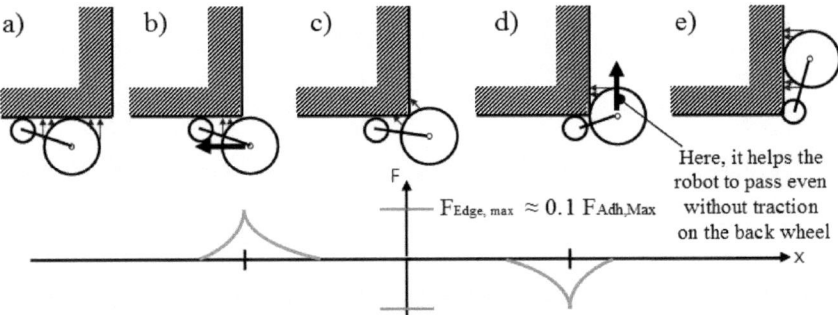

Fig. 4-5: Evolution of the edge pushing force on a magnetic wheel during an outer plane-to-plane-transition; and its influence on the main wheel of a robot with very short wheel axis distance and only front-wheel-traction (taken from the student-report on the micro-Tripod WpW [14])

4.1.2.3 Core parameters for quantifying the difficulty

The difficulty of outer transition can be estimated similar as in inner transition. However, some parameters are less important in this type of obstacle; and some new ones have to be considered as well:

- Angle and thickness in double-edges (often also called ridges or surface flips)

Similar as in inner transitions, also here the angle plays an important role for determining the difficulty – again with acute angles being more severe than obtuse ones – both for assuring enough ground clearance and for dealing with the reduced force caused by saturation effects. However, in most environments only 90° edges occur. Instead of acute angles, the presence of double edges (also called ridges or surface flips, Fig. 4-4-d) is much more common – with the core parameter for describing this type of obstacle being its thickness (b).

- Inclination in respect to gravity

The inclination in respect to gravity again plays an important role, with the transition from ceiling to wall (Fig. 4-4) being the worst case for most robots. However, some types of robots are able to pass outer plane-to-plane transitions in this inclination but fail in another one: The WaalBot [84] cannot move downwards, thus it is of course not able to do transitions in downward-directions. Another example is the CyMag [45], which is able to move from ceiling to wall, but fails in another inclination – from wall to top ground. More details about the mechanisms used in these robots will be shown in Chapter 4.2.2.

- Necessary friction coefficient between robot and surface

While in inner plane-to-plane transitions, a very high friction coefficient between robot and surface is essential for not getting stuck (in some robots, $\mu_{real}>0.5$ is necessary already without considering negative effects of gravity [3]); in outer transition its influence is not as big. In some cases, magnetic wheels with blank steel wheels ($\mu_{real}=0.2$-0.3) even perform better than rubber-covered wheels ($\mu_{real}=0.6$-0.8), as the negative effect of reduced friction gets more than outweighed by the higher adhesion force of using wheels without a rubber gap.

- Curvature radius of the edge ($r_{surface}$)

As already pointed out in the last paragraph and in Fig. 4-4, the curvature of the edge has a very high influence on the force reduction of a magnetic wheel: While on rounded edges (curvature bigger than 10% of the wheel radius) the magnetic force remains almost constant, on very sharp edges the magnetic force can even go down to 20% of its normal value on plain surfaces.

- Presence of burrs or non-magnetic zones

Even the extreme case of a short but almost non-magnetic zone on the edge can occur – on edges that are badly deburred or repaired with resin, copper or another non-magnetic material. If this is the case, the robot can only pass if it is able to completely distribute its necessary adhesion forces among other contact points that are not placed on the non-magnetic zone. Examples of these situations can be found in many applications in power plant inspection.

4.1.2.4 Sub-classes for the classification of outer transitions

To establish comprehensive sub-classes within outer transitions, again the most common requirements from power plant environments are taken into account – as well as specific differences between the most recent robot prototypes. For this reason, only 90°-outer transitions and double edges (with values for "b" coming from real applications) are taken into account. Regarding the inclination, it also distinguishes between the two most difficult ones. The curvature of the edge ("rsurface") is taken into account as well – thus the 90° edges can either be "rounded" (1mm curvature) or "sharp" (<0.1mm radius). The extreme worst case of a burr or a non-magnetic-zone is included as well, as some of the most recent robot prototypes are even able to deal with this type of obstacle.

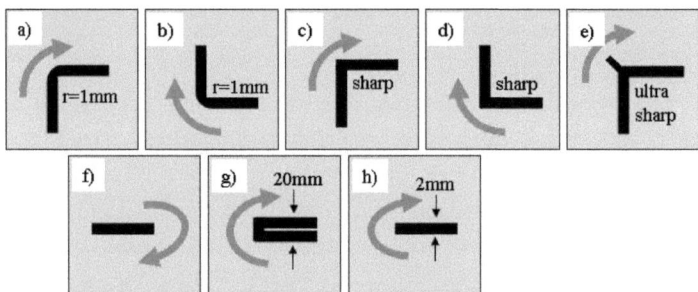

Fig. 4-6: Sub-classes within outer-plane-to-plane-transitions
(a) rounded edge wall to top ground, (b) rounded edge ceiling to wall,
(c) sharp edge wall to top ground, (d) sharp edge ceiling to wall,
(e) edge with an almost non-magnetic burr, (f) double edge down or sideward, (g) thick double edge up, (h) thin double edge up
<u>Origin:</u> (a, b, g) Generator housing, (a-d) Steam chest, (e) Special cases in environments similar to the steam chest, (f, h) Additional functionality offered by some robots, with no link to applications yet

4.1.3 Curvatures

Even if not directly being an "obstacle", curvatures occur in most environments in power plant components – and the ability to deal with them in all variations is a very important characteristic to determine a robot's mobility.

4.1.3.1 Main challenges on axial paths and during turning

In contrast to inner and outer transitions, where it is often enough that a robot is able to pass them at an attack-angle that is more or less perpendicular to the transition line, curvatures should ideally be negotiated at any angle in respect to their axis. While moving on circumferential paths in a curved surface is only problematic for climbing robots on caterpillars or with the adhesion zone in the robot chassis (see Fig. 3-4-c and Fig. 3-10-b), on axial paths also magnetic wheeled robots have to face some challenges: First of all, enough ground clearance has to be guaranteed also in driving direction – especially on convex curvatures. Secondly, magnetic wheels reduce their force when they are tilted on axial paths of a curved surface (Fig. 4-7-a). While the increased risk of slipping caused by this force reduction is normally not a big problem compared to the force reduction on sharp edges (only 50% reduction are reached at approximately 15° tilting angle [5]), the force needed to tilt the wheels while turning the robot on a curved surface should been taken into account when calculating the necessary actuator torque. For some robots with relatively weak motorization in respect to the adhesion force of their wheels (e.g. the robot with passive extra wheels in the structure [12]), the lack of enough traction force for turning in all environments can become a limiting factor for the mobility. In order to deal with the tilting problematic, many robots use structures with passive suspension systems that adapt to curved surfaces. As these suspension systems sometimes increase the mechanical complexity of the robot significantly (Fig. 4-7-b), the MagneBike indeed uses only two main magnetic wheels plus non-magnetic extra wheels for stabilization [5]. This solution allows for designing the robot very compact, but on the other hand needs extra sensors for controlling the active stabilization mechanism and also does not allow for a perfect position of the wheel when the curvatures is convex on one side and concave on the other (see Fig. 4-7-c). A more detailed overview and comparison about the most common types of wheel suspensions in the field of climbing robot is again provided in the corresponding section about obstacle-passing mechanisms (4.2.3.1).

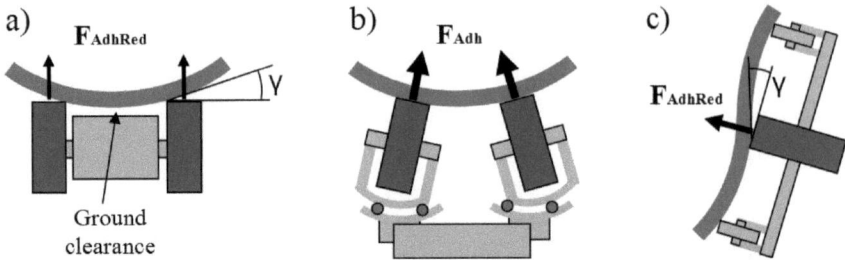

Fig. 4-7: Force reduction on curved surfaces and approaches to reduce its negative effects (a) simple vehicle without suspension, (b) vehicle with a suspension system that rotates cinematically correct (high mechanical complexity), (c) active stabilization of the MagneBike [5] and its limitations on asymmetrically curved surfaces

4.1.3.2 Core parameters and sub-classes for the classification

The most important parameters for classifying the difficulty of a curvature is of course its radius; and if it is concave, convex or a combination of both. As in most applications the robot has to move in any direction in respect to the curvature radius, the worst-case has to be considered – which is normally the axial path and turning into it coming from a circumferential path.

For defining the sub-classes for the obstacle classification, the requirements from typical power plant environments (Fig. 4-8) have been taken – 1.5m concave curvature in generator housings; and 250mm concave curvatures, 500mm convex and combined concave/convex curvature in steam chest environments.

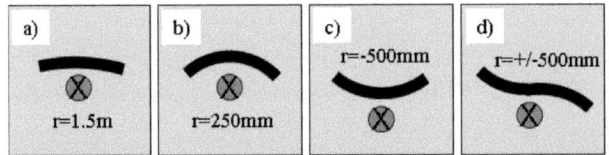

Fig. 4-8: Sub-classes within curvatures (a) concave 1.5m, (b) concave 250mm, (c) convex 500mm, (d) combination +/-500mm; (c-d) Origin: (a) Generator housings, (a-d) Steam chests, while only (b) is absolutely mandatory for reaching all points there

4.1.4 Combinations

In most environments, the above mentioned obstacles normally do not come alone; but very often occur at short distance – often making them even more difficult than just one obstacle alone. Fig. 4-9 shows the typical combined obstacles that can be found in the analyzed power plant environments.

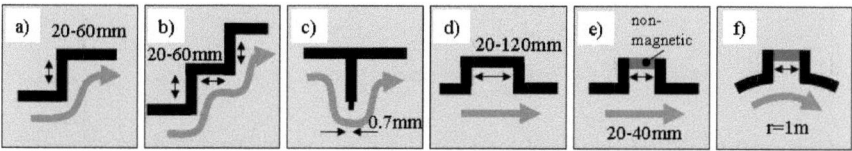

Fig. 4-9: Sub-classes within combined obstacles (a) step, (b) tripple step, (c) thin ridge with saturation on top, (d) long entrance hole, (e) short hole with non-magnetic ground, (f) hole in combination with curvature;
<u>Origin:</u> **(a, b, d) Steam chests with only (a, b) mandatory, (c) Gas storage tanks, (e, f) Generator air gaps, with only (e) absolutely mandatory**

4.1.5 Other hazards

Other hazards sometimes occur as well, limiting the robot's adhesion force or its functionality in general. The ones found in the applications analyzed in this work are listed in Fig. 4-10 – with different impacts on the robot: Thin surfaces significantly limit the adhesion force of big wheels due to saturation effects, while the adhesion force of smaller wheels is not affected significantly, as their magnetic field does not penetrate far into the surface. On the other hand, for rolling on painted surfaces bigger wheels show less problems, as there a large penetration into the surface is favorable to overcome the non-magnetic gap, and even small paint drops are only small bumps instead of big obstacles. In some environments the non-magnetic layer is even so thick, that the surface has to be regarded as completely non-magnetic.

Even worse than paint layers are rusty surfaces – not only limiting the adhesion force due to a non-magnetic layer, but also because small rust particles can get detached when the robot is rolling over them and then afterwards stick on the adhesion zone – where they further decrease the adhesion force and can also damage the robot. While in steam chests, the rust layer is relatively thin and only problematic for robots with very small wheels (~ 15mm radius or smaller), in boilers its thickness is several mm, making the locomotion with magnetic wheeled robots almost impossible.

Another hazard that can be encountered in some environments is sections that are filled with water, which is sometimes at very high pressure (up to 10 bars in boiler tubes). Also extreme heat, electromagnetic waves or radioactivity are hazards that can be encountered in some power plant environments, but have not been analyzed in this work.

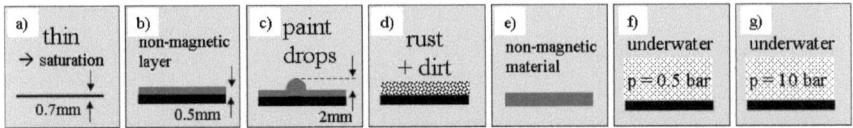

Fig. 4-10: Sub-classes within the other types of hazards (a) thin surfaces with saturation, (b) non-magnetic layers, (c) paint drops, (d) rust + dirt, (e) non-magnetic materials, (f, g) underwater at different pressures
Origin: (a) Gas storage tanks, (b) Generator housing, (b, c) Generator air gap, (d) Boiler, partially also steam chest, (e) Fuel pipes and copper tubes in the cooling system of generators, (f, g) Boiler tubes

4.1.6 Influence of the obstacle size relative to the robot

The difficulty of an obstacle can vary depending on its size relatively to the vehicle size, as it is represented in Fig. 4-11. While small holes, paint drops, non-magnetic layers and/or a small cover of rust usually are no significant limitations for relatively big robots, for very small ones they become severe challenges that sometimes make it impossible to pass. On the other hand, for robots at small size, difficult combined obstacles often get separated into several simple ones, which can be passed much easier. Also the reduction of the adhesion force on edges and thin surfaces normally gets less severe for small robot sizes.

For this reason, it is very important to always provide the size information of each obstacle coming from real environment descriptions – and critically re-ask the question if for a different robot size the challenges remain the same or not. A good example that shows this change of obstacle-difficulty are the test-runs with the first robot for generator housings [12] in the steam chest environment [5]: The tripple-steps and holes that form the worst-case-obstacles for the MagneBike (robot designed for this application, approximately 3-4 times bigger) are split up in a series of well separated inner and outer 90° transitions and can be passed without any problems. On the other hand, small holes of approximately 3mm length (not even mentioned in the specifications) resulted to be almost impossible obstacles for the small wheels of the robot (r=5mm).

Also the rust particles in the test environment caused severe problems for the first prototype of this small robot and even damaged some parts of its structure.

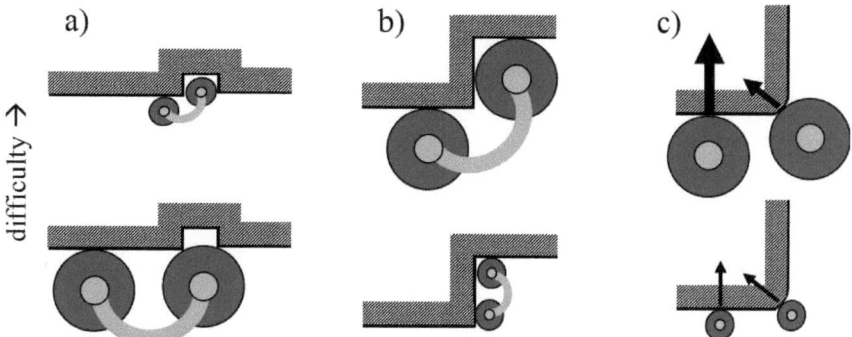

Fig. 4-11: Difficulty of some obstacles depending on their size relative to the robot (a) hole that is just a small bump for a big robot, (b) step that is just a combination of two 90° transitions for a small robot, (c) rounded edge (with some force reduction due to saturation effects) that is just a curved surface (almost no saturation) for a small wheel

4.1.7 Mobility estimation for climbing robots

To sum up, the variety of obstacles and hazards that can be encountered in power plant environments is quite high. Given this high number, and the additional observation that the difficulty of each obstacle normally changes significantly depending on the robot size and the principles that are used for adhesion and locomotion, a simple linear grading scale for the obstacle-difficulty (like for skiing slopes or sport climbing walls) cannot be established.

In contrast, for complete mobility estimation, a robot prototype should ideally be checked against all potential obstacles – or at least the ones that are specified in the application it was developed for. Also the size and the mass of each robot, and optional parameters (e.g. battery + remote control) should be included into such a comparison – both for the robots and for the applications.

A comparison of this type can be found at the end of this chapter– after the detailed description of obstacle passing mechanisms and how they are implemented in different vehicle concepts and prototypes for climbing robots. It uses a matrix with the columns labeled with the applications and prototypes, and the lines labeled with the environment specifications and other important criteria.

4.1. Detailed classification of obstacles

This matrix can be found at the end of chapter 4 and in the appendix, the obstacles and size restrictions that label its lines are drawn in Fig. 4-12. Note that for simplicity reasons the size-, mass- and payload-requirements are not split up into sub-groups that use the values from each application (as it is done for the curvatures). Instead, the quantitative values measured for each prototype are entered in the matrix and compared directly against the required values that are specified in the applications.

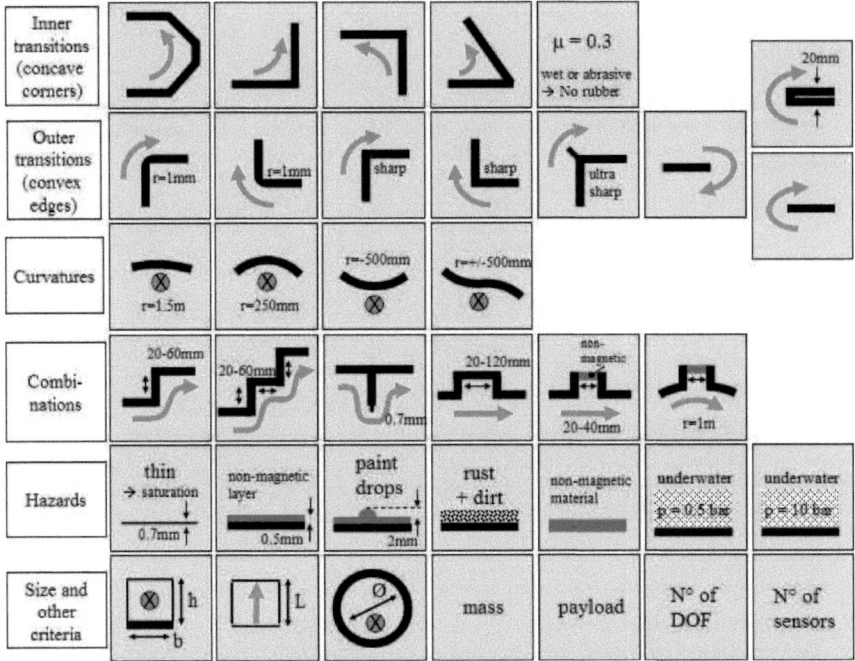

Fig. 4-12: Typical environment specifications for compact climbing robots in power plant inspection, used for the performance evaluation matrix at the end of this chapter

4.2 Mechanisms for increasing the robot mobility

In order to pass the above mentioned obstacles, several mechanisms have been developed or further improved, with the improvements mostly following the goal to simplify the overall robot structure for realizing smaller size, more robust control or higher payload. These mechanisms can be distinguished by the following criteria: Their need for actuation (active or passive), their position in the robot (on the wheels or in the structure), and the obstacle group they are designed for.

In this sub-chapter, the most important mechanisms are described, structured by the obstacle group they are designed for: Inner transitions, outer transitions and combinations (steps, ridges, holes), and other mechanisms to increase mobility.

4.2.1 Mechanisms for inner transitions (concave corners)

As already described in the previous sub-chapter – the main challenge when passing an inner transition, is to deal with the unwanted adhesion force on the old surface (see Fig. 4-1-b). For reducing this unwanted adhesion force, several mechanisms have been proposed – with different goals, motivations and constraints: Low motor torque and required friction coefficient, no need for using an additional motor, universal use on a large variety of angles, small size, etc. The following section provides an overview on the most frequently used mechanisms and concludes with a brief comparison of the most promising ones.

4.2.1.1 Active force reduction

The most intuitive way to reduce the unwanted adhesion force is using an active mechanism in the robot wheel or foot, which allows for changing between high and low adhesion forces. Among these mechanisms, three basic principles can be distinguished – electromagnets, assemblies with moving magnets and lifters.

A relatively simple approach is to use electromagnets instead of permanent magnets and to just turn off the power when the magnetic force has to be canceled. In contrast to the other two principles, this solution has the advantage of extremely fast switching times and that the magnetic force can be cancelled completely. However, the need for continuous power consumption just for maintaining the adhesion (risk of falling in case of a short power shutdown) and the difficulties to combine this principle with small wheels (need for slip rings) make it less favorable for small inspection robots.

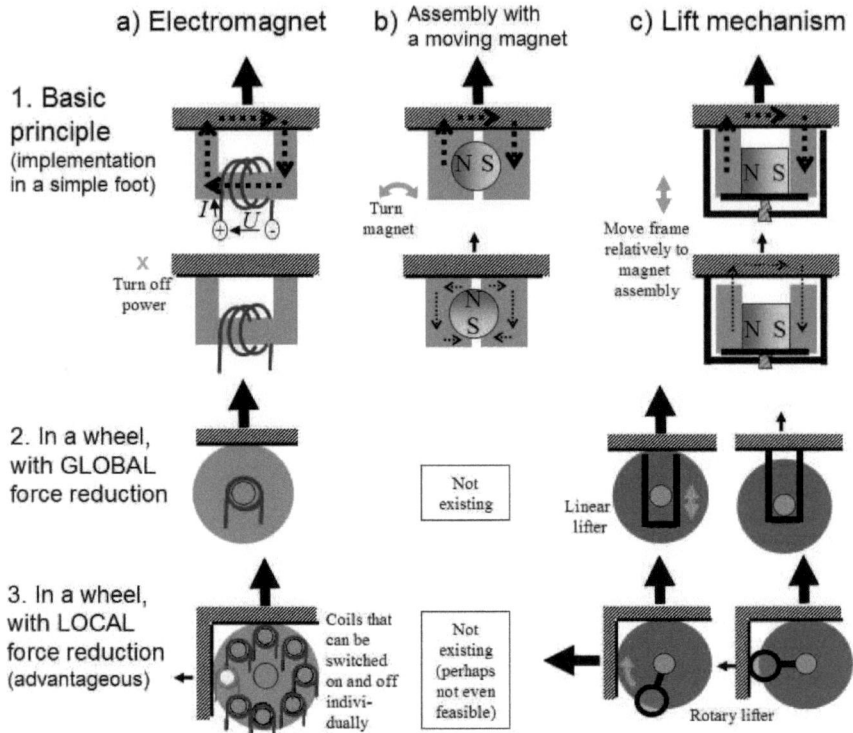

Fig. 4-13: Basic principles for the active reduction of an unwanted magnetic force (a) Use of electromagnets, (b) Assembly with a moving magnet that allows for redirecting the magnetic circuit, (c) Lift mechanisms that lever out the magnetic contact area; (1) sketch of the basic principle, (2) Implementation in a wheel with global force reduction and (3) in a wheel with local force reduction

Another principle that allows for an easy reduction of the magnetic force is to move the magnet within the assembly in order to direct the magnetic circuit either into the surface (high force) or just around the magnet (low force). An example for a simple implementation of this principle can be seen in Fig. 4-13-b. In this design, the circuit-change is done by turning a radial polarized cylinder magnet that lies between two rims with half-circle-holes. More sophisticated designs with lower residual forces (FmagRed), and/or optimizations regarding the torque that is necessary for turning the magnet can be found in recent patents (e.g. the "switchable permanent magnetic device" from Aussie Kids Toy company [52]) or publications from companies that use this technology in the field of industrial robots (e.g. the magnetic grippers by Schmalz GmbH [53]). To our best knowledge, implementations in wheels have not been realized yet.

The third basic principle for active force-reduction is based on moving a non-magnetized part relatively to the magnet assembly in order to lift it away from the surface for a few mm (see Fig. 4-13-c). Caused by the increased air gap in the magnetic circuit, the magnetic force gets reduced significantly. Compared to the force/torque that is required for moving a magnet in an assembly, the necessary force/torque for actuating a lifter is relatively high – which can only achieved by using relatively slow and/or heavy actuators with high reduction gearboxes. On the other hand, the central advantage of this principle is the fact that it can be combined with wheels at relatively low complexity. When moving the lifter not linear but rotary, and with its turning axis coaxial to the wheel axis (Fig. 4-13-c-3), the adhesion force can be decreased locally at any point of the wheel by using just one actuator at a still reasonable torque. As it can be seen in Fig. 4-14, this local force reduction is very advantageous in very small and simple vehicles with only 2 pairs of wheels like the MagneBike [7]. Note that our team – together with ALSTOM - registered a patent [8] on the basic concept of the rotary lifter.

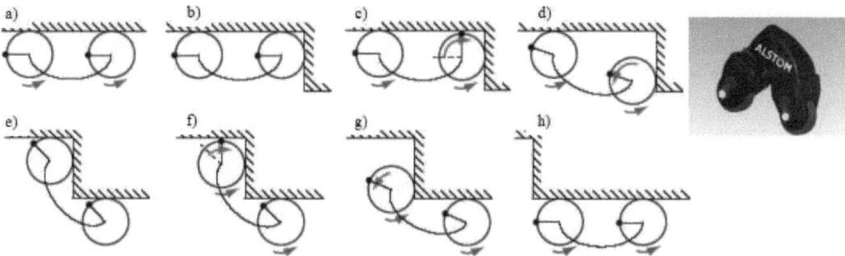

Fig. 4-14: Moving over a step on the ceiling with a vehicle that only has two pairs of wheels and a principle for local force reduction implemented in the wheels (taken from the paper on the MagneBike [7])

Also with the other two technologies (electromagnets and assemblies with moving magnets), some research has been done with the goal to integrate them into wheels. To our best knowledge, only wheels with global force reduction have been successfully realized, and also these wheels were still significantly bigger and more complex than the design with the rotary lifter. For this reason, the use of lift mechanisms – especially the design of the rotary lifter – seems the most promising technology for the active force reduction of magnetic wheels. The other two principles have their main application mainly in the field of legged robots, where the increased switching speed (= time to change between full and reduced adhesion force) and the low mechanical complexity in the "simple-foot"-implementations are significant advantages. More details about legged robots with magnetic adhesion can be found in the corresponding publications (Inchworm [49], MIT Bridge-Inspector [50] and Robinspec [51]).

		Electromagnet	Assembly with a moving magnet	Lift mechanism
	Power consumption in "stand-by"-mode	−	+	+
Important for robots on legs	Remaining residual force (ON is desired)	+	o	−
	Switching speed	+	o	−
Important for robots on wheels	Complexity on wheels with GLOBAL force reduction	o	−	+
	Complexity on wheels with LOCAL force reduction	−	−−	+

Fig. 4-15: Comparison table for the three basic principles for active reduction of the magnetic adhesion force, based on the most important criteria in legged and wheeled climbing robots

4.2.1.2 Extra wheels in the robot structure

Another approach for better passing inner transitions (corners) with magnetic wheeled robots is to avoid the unwanted double-contact in the main wheels by adding non-motorized extra wheels to the structure. By placing the front extra wheels very high and the back extra wheels very low (see Fig. 4-16-b), the minimum friction coefficient between wheel and surface can be reduced to a value below 0.5 – which allows for using very hard and robust rubber coatings on the wheel and/or for driving on wet surfaces. More details about the calculation and optimization of such robot structures can be seen in our paper at IROS09 [12].

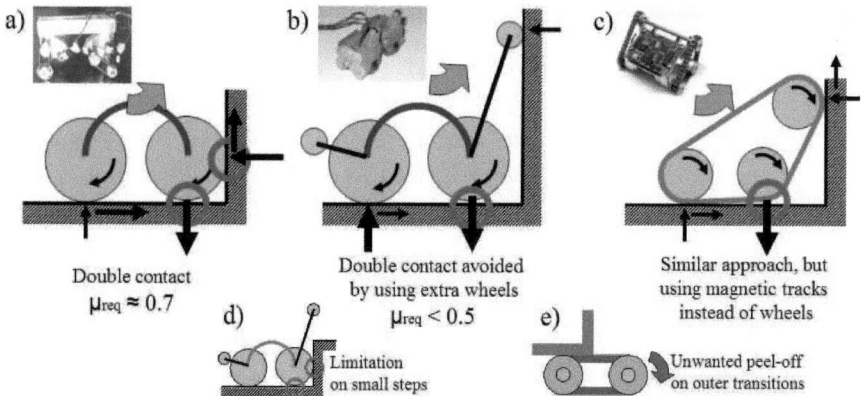

Fig. 4-16: Comparison of basic magnetic wheeled robot structures in inner transitions and reduction of μ_{req} by using extra wheels in the structure (a) First ETH test prototype [3], (b) Robot prototype for generator housings [12], (c) Similar structure in the TriPillar [39], (d) Limitation of high front wheels on small steps, (e) Unwanted peel-off with caterpillars

The same concept of using high front wheels for facilitating inner transitions is also implemented in the TriPillar ([39] and Fig. 4-16-c, realized by our partners at EPFL). In contrast to the ETH-robot for generator housings, it uses a combination of magnetized caterpillars and structure magnets for adhesion – with the goal to achieve a better friction coefficient towards the surface and to move in environments with small holes and gaps. Similar to the ETH-robot, it is also able to pass inner transitions, but fails on outer transitions (edges) and convex curved surfaces as the caterpillars get peeled off there (see Fig. 4-16-e and chapter 4.2.2).

When comparing the vehicles with extra wheels in the robot structure to vehicles with active mechanisms for force reduction (e.g. the MagneBike), their main advantage is the fact that no extra actuators are necessary for passing inner transitions – allowing for very small and simple devices. However, the design also shows some limitations: On small steps (lower than the front wheel) the extra wheels cannot avoid the unwanted double contact on the main wheels (see Fig. 4-16-d. Also the required friction coefficient is still higher than for just normal driving and passing the corners by using a mechanism for active force reduction.

4.2.1.3 Passive lift mechanisms directly on or close to the wheel

For avoiding the above mentioned problem on small steps, other passive lift mechanisms have been developed and analyzed as well in the context of this work - with the contact points for levering out the main wheel at lower positions: Passive-front-arm, Tri-Wedge and Wheel-parallel-to-the-wheel (WpW).

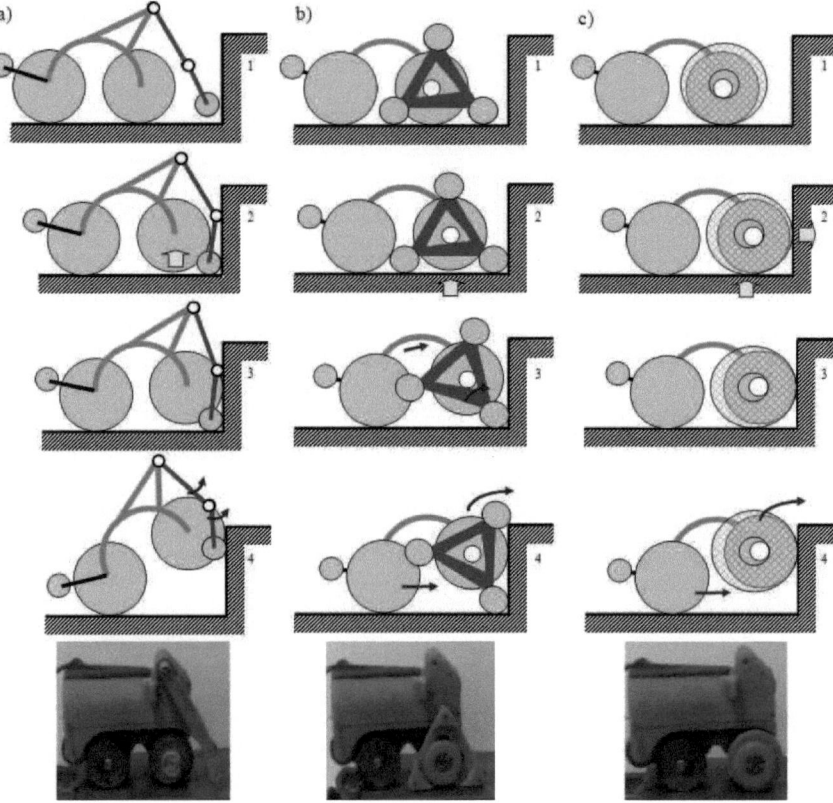

Fig. 4-17: Passive lift mechanisms with low contact points, for better performance on small steps (a) Passive front arm, (b) Tri-Wedge, (c) Wheel-parallel-to-wheel (WpW), and motion sequences for passing a step (1-4)

In the "passive-front-arm"-structure (see Fig. 4-17-a), an extra wheel is mounted on a spring-loaded double arm that rolls in front of the robot. When this extra wheel gets in contact to the new wall after an inner transition (1), it gets pushed down into the corner and levers out the main wheel (2). Getting pushed by the

4. Obstacle-passing with compact climbing robots

back wheel, the front main wheel gets again into contact with the new surface after the transition (3). By designing the arm in two parts and wisely choosing the lengths of both segments and the spring stiffness of the joints, the extra wheel is then able to move back into its initial position (4). Similar to the design described in the last sub-section, also here a low extra wheel at the back facilitates the robot passing at the end of transitions and the extra wheels are made out of magnetic material for not losing contact in transitions from wall to ceiling.

Even if this design performed relatively well in the tests, it shows some disadvantages towards the other alternatives that are described in the next paragraphs: Relatively high mechanical complexity (2 spring-loaded DOF), limitation to only the front wheel and need to push with the back wheel (only works on robots with all-wheel-traction). For solving these problems, two improved designs have been derived from this idea – Tri-Wedge and Wheel-parallel-to-the-wheel.

In the Tri-Wedge (see Fig. 4-17-b), the front wheel rolls onto a wedge for getting lifted before it gets into double contact at the corner. To realize this wedge-effect only there, a frame with three extra wheels passively rolls parallel to the main wheels on an axis with smaller diameter than the main wheel (1). When approaching an inner transition, the frame gets stuck in the corner and the axis with smaller diameter rolls on the wedge (2) – levering out the main wheel on the old surface (3). After having rolled to the new surface, the wedge gets then free again and continues rolling as before (4).

Similar to the first mechanism, also this one performed well in the tests but still shows a relatively high mechanical complexity and could get stuck when rolling on uneven surfaces. Using the same basic idea but in a much simpler implementation, the Wheel-parallel-to-the-wheel (WpW, see Fig. 4-17-c) has been developed. Similar to the Tri-Wedge, also in the WpW a frame rolls passively to the main wheel and is guided by an axis with smaller diameter that then serves as a wedge when levering out the wheel off the old surface. In contrast to the above mentioned mechanism, the frame in this design is just formed by a simple disc out of non-magnetic material. Tests showed that by choosing the diameter of the disc only a little bit bigger (around 10%) than the main wheel, the contact can always be assured even in unfavorable inclinations – as the maximum gap between wheel and surface still remains small enough for providing the necessary adhesion force to prevent the robot from falling (see Fig. 4-19). In addition to the significantly lower complexity compared to the other two mechanisms, tests also showed that the WpW is even powerful enough to allow for inner transitions with traction on only one wheel pair [**Error! Reference source not found.**] (was not possible with the other two mechanisms). For this reason, the wheel-parallel-to-wheel can clearly be regarded as the most promising principle within the passive lift mechanisms. It is currently under more detailed investigation at ALSTOM, for deciding about a possible patent application [**Error! Reference source not found.**] and future robots where it can be applied.

4.2.1.4 Dual magnetic wheel and multiple wheel-inside-the-wheel

Another principle that is very similar to the Wheel-parallel-to-the-wheel (WpW) is the "dual magnetic wheel" [40] or also called "wheel-inside-the-wheel". While the sequence of passing an inner transition looks quite similar at first glance, the main difference lies in the material of the disc and where it rolls on: In the WpW, the disc is out of non-magnetic material and only comes into action during the transition, where it serves as a wedge for levering out the wheel on the old surface. In contrast to this design, in the dual magnetic wheel the main magnetic wheel always rolls in the disc (or outer wheel) that is made out of ferromagnetic steel and lies between main wheel and surface. Compared to the WpW, this design brings two main disadvantages: Reduced adhesion force also in normal rolling conditions (2 times line contact with saturation effects); and slightly more complex to realize at small size (need to connect the steel discs for axially guiding them, but with non-magnetic material to avoid a shortcut in the magnetic flux: challenging to manufacture all this perfectly coaxial – especially at small size).

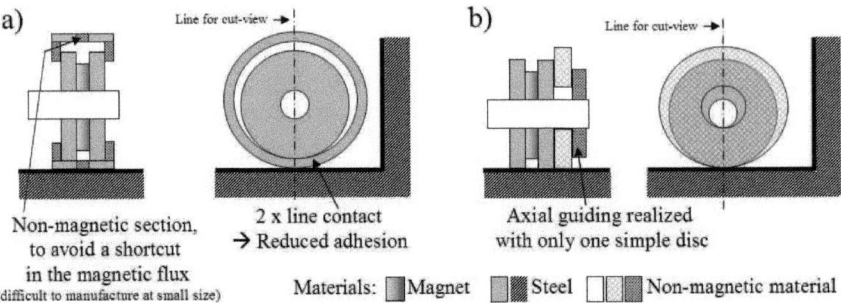

Fig. 4-18: Schematic cut- and side-view of the dual magnetic wheel (a) and the wheel-parallel-to-the-wheel (b), showing the main disadvantages of the dual magnetic wheel in comparison towards the WpW

4. Obstacle-passing with compact climbing robots

Besides the WpW, also another mechanism has been realized that has a high similarity to the dual magnetic wheel – the "multiple wheel-inside-the-wheel" (see Fig. 4-19-b). The main motivation for this concept was to improve the adhesion force when passing inner transitions from wall to ceiling. As it can be seen in Fig. 4-19-a, the adhesion force of a dual magnetic wheel can get significantly reduced in the middle of the transition, especially if the diameter of the outer wheel is large compared to the inner one. By replacing the big inner wheel by several small inner wheels, not only the mass can be reduced significantly; but also the adhesion force distribution during the transition gets smoother. This idea – originally generated by Christophe Groux in his bachelor thesis at EPFL [42] – has been detailed in two generations of prototypes, but to our best knowledge never resulted in a reliable robot with advantages over similar designs – mainly because of the mechanical complexity that is necessary for motorizing all inner wheels.

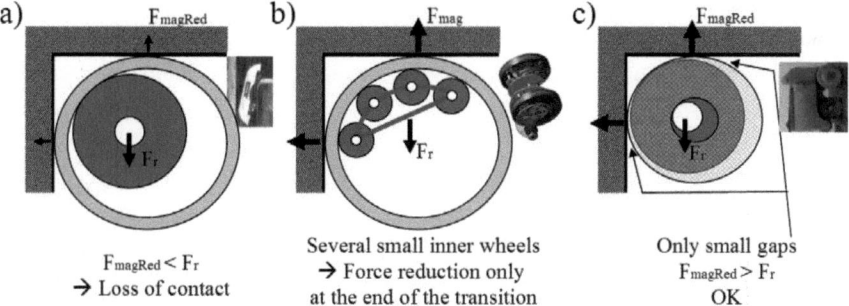

Fig. 4-19: **Approaches to decrease the risk of losing contact when passively passing inner transitions from wall to ceiling (a) Problem of reduced magnetic force when using a dual magnetic wheel with too large disc diameter, (b) Multiple wheel-inside-the-wheel-concept [42], (c) Design with only small diameter differences, in the WpW-concept [10]**

Note that the above mentioned problem of losing contact during the critical transitions is less critical in the WpW-concept, as it is relatively easy to design the outer discs and main wheels with only very small diameter differences – which makes the force reduction only relatively small (comparable to the force-reduction on outer transition) and thus is not critical any more (see Fig. 4-19-c).

4.2.1.5 Moving magnetic adhesion zone

Another approach for passing corners is to replace the magnetic wheels by a combination of normal wheels and movable magnetic adhesion zones that can be turned into the new direction during the inner transition (see Fig. 4-20). This concept was originally invented by Osaka Gas as a potential alternative to the dual magnetic wheel [41], further improved by our partners at EPFL, and analyzed more detailed in several student projects [43-45].

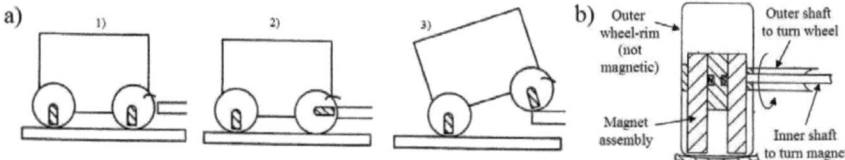

Fig. 4-20: Corner-passing by using a moving magnetic adhesion zone close to the wheel (a) Motion sequence for passing a step, (b) Cut-view through a wheel (pictures taken from the patent by Osaka Gas [41])

Up to now, three main design alternatives have been realized in prototypes: Actively moving the adhesion zone by using an extra motor (SpokeHeel, [43]), passively moving it with a mechanism that is relatively big and mechanically complex compared to previously described mechanisms [44], and a special vehicle structure that allows for a very simple and compact design but is not able to pass edges in all inclinations (CyMag, [45]). A short overview on these prototypes and their main limitations is represented in Fig. 4-21.

Compared to the prototypes using one of the previously described mechanisms (e.g. the WpW), robots with a moving adhesion zone either have a significantly lower mobility on specific obstacles (outer transitions and curved surfaces – see next sections) and/or need a system complexity that is significantly higher than comparable mechanisms with similar mobility. To our best knowledge, plausible solutions for solving these drawbacks have not been found yet.

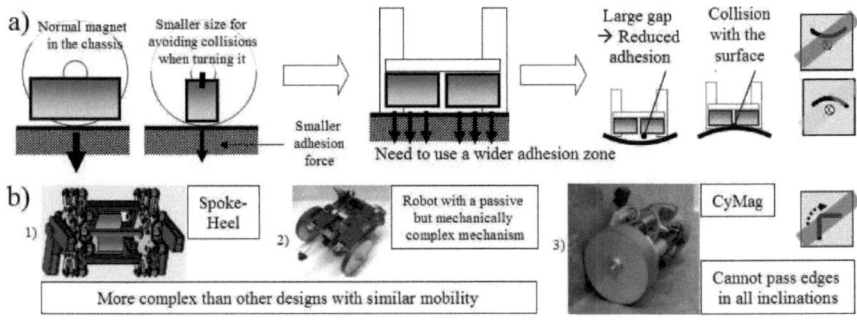

Fig. 4-21: Main limitations of the robots with a moving adhesion zone (a) Problems on curved surfaces, (b) Prototypes realized by our partners at EPFL [43-45] and their main drawbacks towards other designs

4.2.1.6 Alternatives to rolling with round wheels

The use of alternative roll-legged locomotion principles instead of wheels can allow for corner-passing ability as well: As it was shown in recent video-contributions at ICRA09, current robots rolling on whegs or elastic caterpillars are also able to pass inner transitions – the WaalBot [84], the Gel-Type-Sticky Mobile inspector [85] and the TankBot [87]. Note that all these robots use non-magnetic adhesion principles, and for this reason can also run on surfaces that are not made out of steel. What can be observed when analyzing the videos and the attached papers in more detail is the following: The robots on whegs never move downwards, while the robot on elastic caterpillars is not shown on outer transitions from ceiling to wall. Given the design restrictions of these prototypes (whegs only work in one direction and the tails are not adhesive, peeling-effect of caterpillars on outer transitions), it is very likely that these experiments have been "actively forgotten" in the videos – which basically means that these types of robots are not mobile enough (yet) for the challenging tasks in power plant inspection that were analyzed in the context of this thesis.

Inspired by these designs, also implementations with magnetic adhesion have been realized: The simplest design variation of this type is just a magnetic wheel with the steel-rims (round in the standard design) replaced by screw nuts (hexagonal). Tests with a very simple prototype showed that with this design it is possible to pass inner transition even without a rubber cover on the wheel (Rotating magnetic cam-disc [15]). The main disadvantage of this mechanism is however that by rolling on such type of "whegs", the motion is very inhomogeneous, with negative effects on sensors and electronics on the robot. A more complex magnetic implementation developed by Ben-Gurion University [46] uses several magnets that can move in radial direction and are spring-loaded for returning into a neutral position. This design allows for passively rolling through inner transitions by using the peeling effect – similar to robots on caterpillars. However, this peeling effect is also showing its limitation on outer transitions from ceiling to wall, where the robot cannot pass to our best knowledge.

A brief overview on these alternative roll-legged locomotion principles with passive-corner-passing-capability and their main limitations are shown in Fig. 4-22.

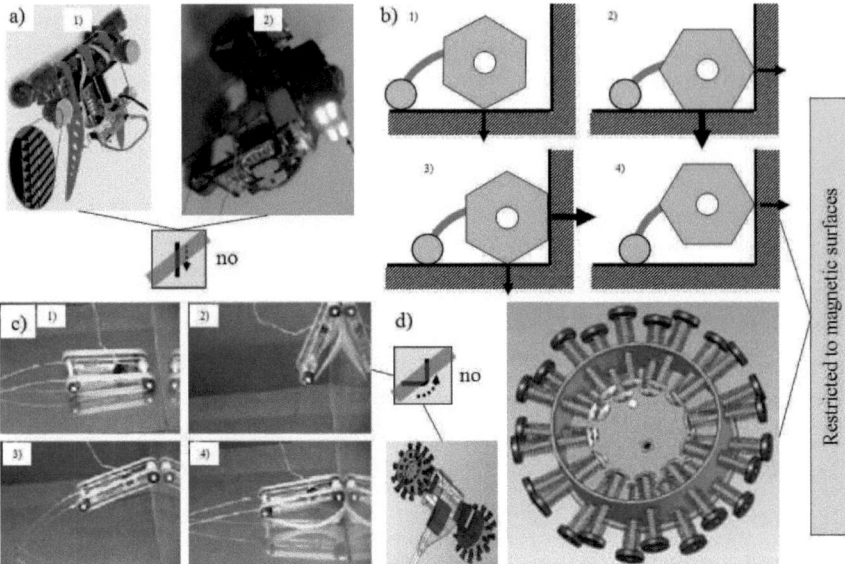

Fig. 4-22: Alternative roll-legged locomotion principles with the ability to passively roll through corners, and their most important limitations (a) Whegs, implemented in the WaalBot [84] and the Gel-Type-Sticky Mobile Inspector [85]), (b) Rotating magnetic cam disc [15], (c) Elastic caterpillars in the TankBot [87], (d) Special wheel with magnets that can move in radial direction, by Ben-Gurion University [46]

4.2.1.7 Overview and comparison

To sum up, in total 6 principles for facilitating inner transitions with compact climbing robots have been presented in this sub-chapter. For the mechanical implementation of these principles, approximately 2-4 design alternatives per principle have been analyzed and described in detail, showing specific advantages and limitations for the mobility of compact climbing robots.

A comprehensive comparison of the most important design alternatives is represented in Fig. 4-23 – highlighting the most important advantages and limitations of each mechanism or mechanism group. It can be observed, that no concept is outstanding in all criteria and without any disadvantages. For this reason, the choice of a concept should always depend on the application and the environment where the robot is planned to be applied.

4. Obstacle-passing with compact climbing robots

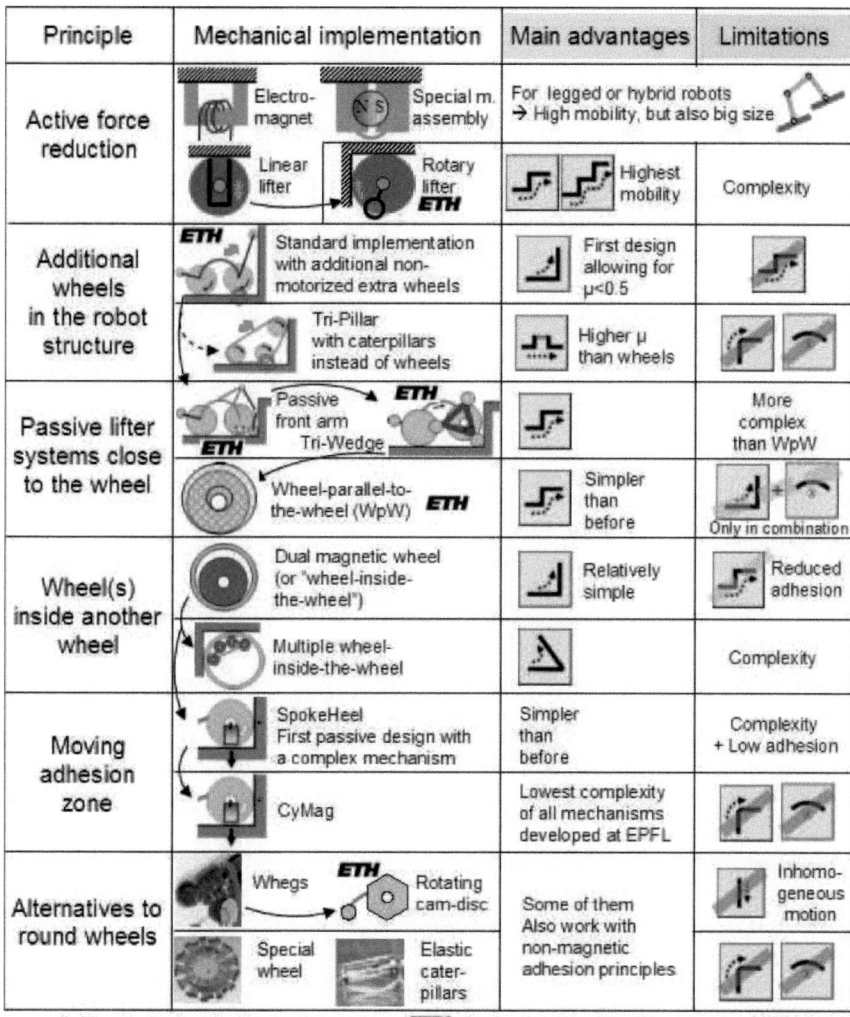

Fig. 4-23: Overview on the basic principles for facilitating inner transitions with compact climbing robots and the most important concepts for their mechanical implementation – showing both the main advantages of current robot designs and their most severe limitations

However, some observations can be concluded:

- Global active force reduction with linear lifters, assemblies with a moving magnet, electromagnets or other active adhesion principles is only favorable in robots with active joints in the structure (higher number of DOF), which leads both to a very high mobility but also to a size that is significantly bigger than in other solutions. Such robots, usually realized in biped-configuration [49, 50] are useful in environments with extremely difficult obstacles but not very hard restrictions regarding size, mass and/or speed.

- For robots with "medium" size and complexity (=around 100-200mm and 3-5 DOF), active force reduction by using a rotary lift mechanism is a promising approach. An example for such a robot is the MagneBike [7], which combines high mobility on inner transitions, curvatures and combined obstacles at a complexity that is still reasonable at this size.

- For robots at smaller size, passive principles should be preferred. Within these principles, the Wheel-parallel-to-the-wheel (WpW) [10] is both the mechanically simplest and the one where the disadvantages seem the least severe. Concerning the developments in our team, it is the last and the most advanced generation (Structure with extra-wheels → Passive front-arm → Tri-Wedge → WpW).

- A similar evolution of corner-passing-mechanisms has also been performed by our partners at EPFL, leading to quite remarkable results with the most recent vehicles using a rotating adhesion zone [43-45]. However, the limitations on outer transitions and curved surfaces that come along with these principles are not solved yet – which significantly reduces their applicability in compact robots for power plant inspection.

- Similar to the principles with the rotating adhesion zone, also the limitations for the concepts that use alternatives to round wheels [15, 46, 84, 85 and 87] still remain relatively severe and thus do not seem promising for power plant inspection yet. However, some of them can also be applied to robots that use non-magnetic adhesion principles [84, 85 and 87]. For this reason, they could once become promising as well – especially for inspection and cleaning tasks in non-magnetic environments, in military applications and/or in search + rescue-operations.

4.2.2 Mechanisms for outer transitions + combined obstacles

As already explained at the beginning of this chapter (4.1.2) and again visualized in Fig. 4-24, outer transitions with rounded edges that still allow for a reasonable adhesion force can be passed by simple magnetic wheeled robots with enough ground clearance, while other roll-legged configurations (robots rolling on caterpillars or whegs, or with the adhesion zone in the chassis) already fail on this relatively simple type of outer transition.

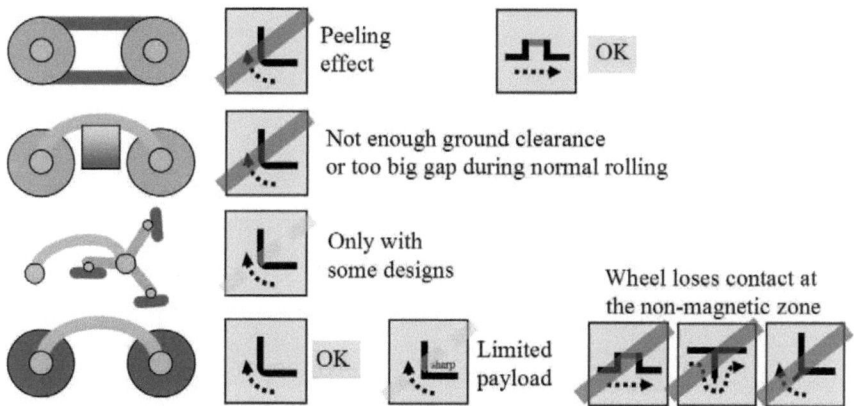

Fig. 4-24: Performance on outer transitions and combined obstacles, for the 4 basic roll-legged configurations in compact climbing robots: caterpillars, adhesion zone in the robot chassis, whegs, magnetic wheels

However, on sharp edges with a large reduction of the adhesion force, the payload of such simple designs with only 2 pairs of wheels gets significantly reduced. On more difficult obstacles with zones that can be regarded as almost non-magnetic (very sharp-edges, thin ridges or holes with non-magnetic ground), such simple vehicle structures are not even able to pass at all – if there is no mechanism to move back the wheel that has lost the contact to the surface.

4.2.2.1 Vehicle structures with active joints in the chassis

For moving back such a wheel after having lost contact in difficult outer transitions or combined obstacles, the most universal solution is to use a robot structure with more than two pairs of wheels and active joints in the chassis. This approach has been realized in many different robot concepts and prototypes.

Within these concepts, several variation parameters exist – having a significant influence on the robot's mobility and its complexity. For better understanding these influences, the most important ones are listed below and discussed briefly:

- Total number of wheels (or wheel pairs): In most robots of this type, the number of wheel pairs is just three, driven by the goal to keep the overall number of components and actuators as low as possible. However, three-wheeled combinations normally require more complex solutions for the adhesion and/or the motorization of the wheels. For this reason, structures with four wheel pairs sometimes result in lower complexity than three-wheeled structures (e.g. [1] vs. [48]).

- Adhesion type of each wheel pair: Another main criterion for distinguishing such vehicle structures is the adhesion type implemented in each wheel pair. In the simplest possible configuration, only the central wheels are magnetic, while the outer ones are just non-magnetic. Other designs use magnetic wheels on all axes; and the most complex designs even include a lifter or another mechanism for active force reduction (see chapter 4.2.1.1) on the wheels [48].

- Number of wheel pairs that are motorized: The number of motorized wheels is another variation parameter. While in the simplest possible structures only one wheel pair is motorized [14], also many robots exist where all wheels are motorized for achieving maximum traction in all situations.

- Type of joint: In most robots, rotary joints are used as this form of linkage allows for higher universality on both inner and outer transitions, while robots with linear motion are normally limited to steps or ridges. However, linear motion also shows some advantages, especially for providing high forces with relatively light and compact actuators (e.g. the spindle motors used in the robot for gas tanks [1]).

- Number of joints: Also the number of joints can vary – normally in the range between 1 and 3 – with high mobility in the case of many joints but also paying the price for increased mass and complexity.

Keeping these main variation parameters in mind, a selection of common vehicle structures with active joints is listed and compared in Fig. 4-25.

4. Obstacle-passing with compact climbing robots

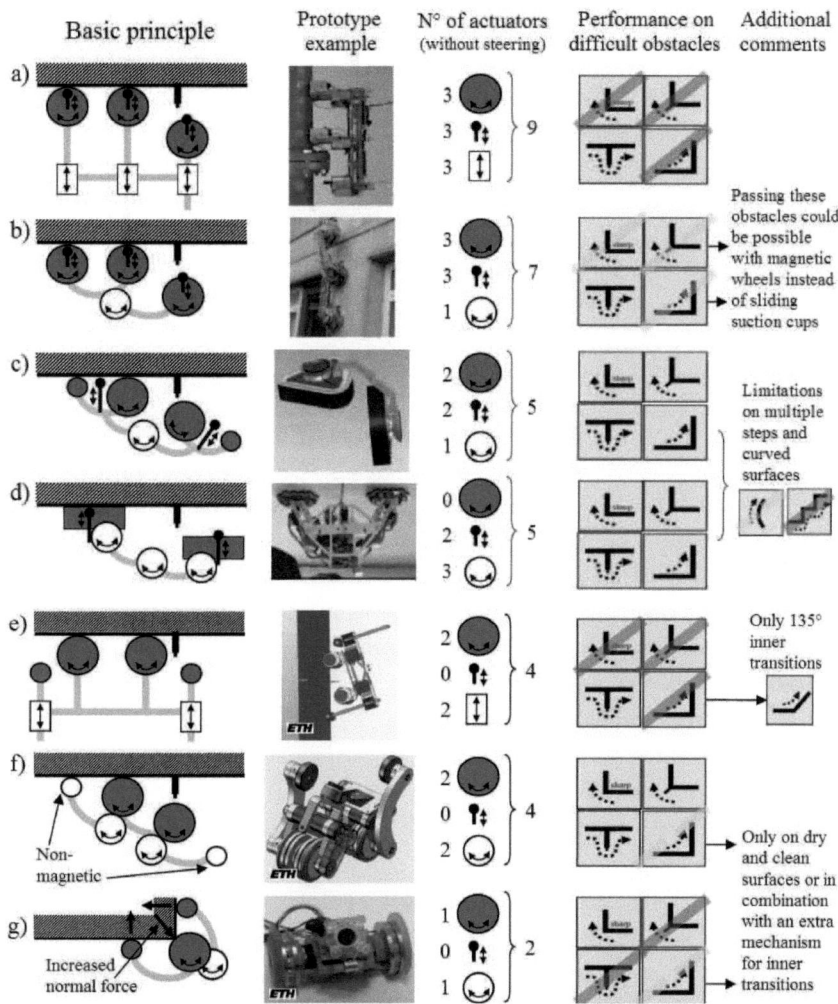

Fig. 4-25: Vehicle structures with active joints for outer transitions and combined obstacles, with prototype examples, the number of DOF for roughly estimating the complexity and a rough performance evaluation on difficult obstacles. Referred prototypes: (a) PIR [48], (b) Alicia^3 [80], (c) City-Climber [82], (d) Roma II [75], (e) Concept for gas tank inspection [1], (f) MagneBike PCAE [9], (g) Micro-Tripod WpW [14]

The most intuitive way for such a structure is to couple 3 identical modules in an active structure – with each module consisting of a motorized wheel (pair) and an active mechanism for force reduction (or an active adhesion principle) – as shown in Fig. 4-25-a and b. The active structure can either be realized with one, two or three joints – and with the joints being either rotary or linear. Prototype examples for such structures are the Alicia^3 [80] or the PIR [48]. Given the relatively high number of actuators that is necessary for such structures (7 or up to 9, without steering) and the observation that the obstacle-passing-capability is even lower than in most other solutions, these concepts do not look very promising for realizing very compact climbing robots.

A simpler structure with only two modules and one rotary joint in between has been realized in the City-Climber-prototype II [82]. In contrast to the Alicia-modules, the City-Climber-modules can not only transmit normal forces towards the surface but also a momentum. This increased stability allows for using only two modules instead of three, which significantly decreases the overall size and system complexity. In an implementation with magnetic wheels (not realized yet, but conceptually drawn in Fig. 4-25-c) this design would correspond to modules with a motorized wheel pair, a non-motorized castor wheel and a lift-off-mechanism for the entire module. This structure is very similar to a biped structure (Fig. 4-25-d, example prototype: Roma II [75]) – with the only difference that the motorization of the wheels is replaced by joints in the structure there. Both types allow for a very high mobility on sharp outer transitions, ridges, small steps and inner transitions. However, the relatively large surface of the drive modules or the big feet brings limitations on multiple steps, curvatures or other geometries in the range of the module's size – at least in the current prototypes that mainly use vacuum suction for adhesion. As already described in the chapter about basic locomotion principles (3.2.4), biped locomotion additionally brings the disadvantages of slow speed, complex control and discontinuous motion – which is not favorable for some types of NDT-measurements.

An even simpler vehicle structure that allows for passing ridges and 135° inner transitions has been proposed by our team in the context of the first project on gas tank inspection [1]. In this vehicle structure (see Fig. 4-25-e) only the inner wheels are motorized, while the outer wheels are passive, but can move in linear bearings perpendicular to the surface. By choosing extra-strong stepper motors with integrated spindles (300N force at only 300g actuator mass) for powering the linear motion, the use of additional actuators for active force reduction can be avoided – which leads to a further simplification of the vehicle structure.

As already explained at the beginning of this section, linear actuators allow for very high forces, but their use is less universal when both inner and outer transitions have to be passed. For also dealing with obstacles of this type without increasing the total number of actuators, the "active-edge-extension" in the MagneBike PCAE (PassiveCorner-ActiveEdge) [9] uses almost the same basic vehicle concept – but with non-magnetic extra-wheels and rotary joints instead of linear ones. As rotary joints at reasonable size and mass normally only provide enough torque for dealing with the gravity of the robot, but not for lifting off the strong magnetic wheels (requires around 3-4 times more force/torque), this design takes advantage of two observations: Actively lifting off a wheel before a sharp edge or ridge is not necessary, as it anyway loses contact when it rolls over the zone of low adhesion. And for moving back a wheel that has lost contact, also non-magnetic outer wheels can be used. With the configuration drawn in Fig. 4-25-f, a magnetic wheeled climbing robot is able to move on all types of outer transitions and most combined obstacles at only 4 active DOF (without steering). However, for passing inner transitions also on wet or dirty surfaces, such a vehicle structure still needs an additional mechanism as described in chapter 4.2.1 (e.g. the WpW).

The simplest variant within active structures for improving the mobility on sharp outer transitions is the one implemented in the Micro-Tripod WpW [14]: With only one motorized wheel pair, two passive wheels and an active rotary joint in the structure, this mechanism only helps to increase the normal force on the main wheel during the worst case in sharp outer transitions (Fig. 4-25-g) but does not allow for passing thin ridges or other difficult combined obstacles. However, its unbeatable simplicity allows for realizing extremely small vehicles in the range of only a few cm. Tests with such a vehicle showed that the payload capability of these vehicles can be increased more than factor 2 in comparison to designs that just try to passively roll over sharp edges (see Fig. 5-28).

4.2.2.2 Passive mechanisms for outer transitions and combined obstacles

Also for outer transitions and combined obstacles, passive mechanisms have been investigated – however their number is still relatively small compared to the active vehicle structures or the passive mechanisms for inner transitions.

In environments where small holes or gaps have to be passed regularly, the use of caterpillars instead of wheels is the most common approach (see Fig. 4-26-a). If additionally the allowed height for such vehicles is restricted to only a few mm (e.g. in generator air gaps), a more compact alternative to caterpillars is to use units with several rows of magnetic wheels. More details about the design of such units and the specific design challenges of robots at extremely low height can be found in the case study about the inspection of generator air gaps (chapter 5.4.1) and the corresponding publications [16-21].

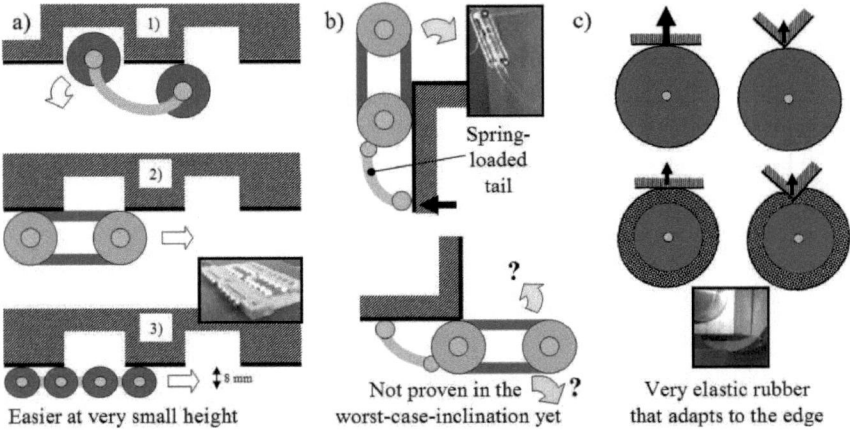

Fig. 4-26: Passive mechanisms for outer transitions
(a) Gaps/holes: (1) Problem for robots with only 2 wheel pairs, (2) Caterpillar structure that is able to pass, (3) Robot at extremely small height with several wheels in serial instead of caterpillars (Air-Gap-Crawler[16]),
(b) Approach to solve the peeling problem on caterpillars on edges by using a spring-loaded tail (TankBot [87]),
(c) Special wheel that better adapts to the edge thanks to a very elastic rubber (study in a semester work at our lab [47])

For improving the performance on edges, one approach is to use a spring-loaded tail at the back of the robot, as it is done in the TankBot [87]. With this tail, the robot is able to pass edges from a vertical wall to a flat surface on the top (see Fig. 4-26-b), a transition that cannot be solved with most other robots running on adhesive caterpillars (e.g. the TriPillar [39]). However, the functionality also on the worst-case inclination (ceiling to wall) has not been proven yet. Prototypes that combine wheeled locomotion with such passive tails do not exist yet – to our best knowledge.

In the field of wheeled robots, only one work on passive mechanisms for edges is known to our team - a student project on experiments with very flexible magnetic wheels, which has been done at our lab in spring 2009 [47]. The basic idea behind this approach is to use a very thick and elastic rubber tire, that gets deformed stronger on edges than on normal surfaces and thus allows for fewer differences between maximum adhesion force on the plain surface and reduced adhesion force on the sharp edge. For improving the magnetic conductivity of the rubber, also some experiments with mixtures that include some steel dust in the molding-mass have been performed. What could be concluded from this project was the following: Even with the best tested values for tire thickness and rubber material, the performance on sharp edges was equal or even lower than with standard wheels. However, in inner transitions the wheels were able to pass more easily than normal ones – using a similar effect as in the special magnetic wheel that was developed by Ben-Gurion University (see Fig. 4-22-c and [46]).

4.2.2.3 Overview and brief summary

To sum up, for the basic forms of outer transitions and steps (rounded edges with few saturation and steps), wheeled locomotion with all-wheel-traction is normally enough and no special mechanisms are needed. In contrast to inner transitions, the approach to use passive mechanisms on more difficult obstacles in this group (extra sharp edges, or ridges) did not show promising results (yet) – with the only exception of caterpillars or several wheels in serial for passing the gaps in generator stators. In the field of active vehicle structures, many different concepts exist – with the following ones showing the best compromises between simplicity and mobility increase:

- Front-arm-extension (see Fig. 4-25-g): Lowest complexity but relatively small mobility increase (higher adhesion force on sharp edges)

- Edge-extension for the MagneBike (see Fig. 4-25-f): Relatively low complexity, second-highest mobility on transitions and no limitations on curved surfaces.

- Structures with 2 movable and detachable adhesion modules; and classical biped-structures (see Fig. 4-25-c and d): Still reasonable complexity, highest mobility in transitions, but limitations on multiple small steps or curved surfaces.

4.2.3 Other mechanisms for increasing the mobility

Beside the mechanisms for inner and outer transitions, also mechanisms for better adapting to curved surfaces, for changing the direction of motion, or for passing very difficult obstacles close to the entrance can be implemented. .

4.2.3.1 Wheel suspension mechanisms for better adapting to curved surfaces

As already explained in the introduction chapter about curvatures (4.1.3.1), the adhesion force of a magnetic wheel gets decreased when it is tilted – caused by the increased air gap on one of the rims (see Fig. 4-27-a).

For avoiding this force reduction, many robots use suspension mechanisms around the wheels that allow for passively turning the wheels plus their motors along an axis parallel to their rolling direction. With this additional passive degree of freedom, it can be assured that always both rims of each magnetic wheel stay in contact to the surface. However, such suspension systems also bring disadvantages: If realized with only one simple rotational joint (see Fig. 4-27-b), the wheels slip in lateral direction (Δy) when the robot is turning on a curved surface. This slip is not only unfavorable for measuring the odometry, in case of relatively thin wheels and a high rotation point, the wheels can even get tilted unwanted. For this reason, usually a more complex mechanism is used, where the suspension axis goes directly through the contact point between wheel and surface. Such a suspension can either be realized with an articulated bar structure or with two slide bearings (see Fig. 4-27-c) – with both possibilities needing lots of space.

In order to build more compact, in the MagneBike [7] another approach was followed: The magnetic force is provided by just one main magnetic wheel in the center, while the stabilization on traversing paths along vertical walls is done actively – with two non-magnetic wheels on the side (see Fig. 4-27-d). By also using these stabilizer wheels as rotary lifters for active force reduction on inner transitions (see chapter 4.2.1.1), the same actuators can be used for two functions. However, this structure also has some drawbacks: First of all, the position of the stabilizer wheels needs to be controlled permanently – for not lifting off the main wheels when it is not desired. This permanent need for control does not only slow down the robot speed, it also requires additional force sensors on all 4 stabilizer arms. Finding an appropriate solution for these sensors and the transmission of their signals to their robot resulted to be the most challenging and time-consuming task within the detailed design and the industrialization of this robot.

Additionally, the tilting of the wheels also cannot be avoided in environments, where the surface is curved in two directions.

Driven by these drawbacks, the most recent versions of magnetic wheeled robots (MagneBike PCAE [9] and Micro-Tripod WpW [14]) run without any mechanism for curvature adaptation and take into account that on curved surfaces the adhesion force can get decreased down to 50% of the normal value.

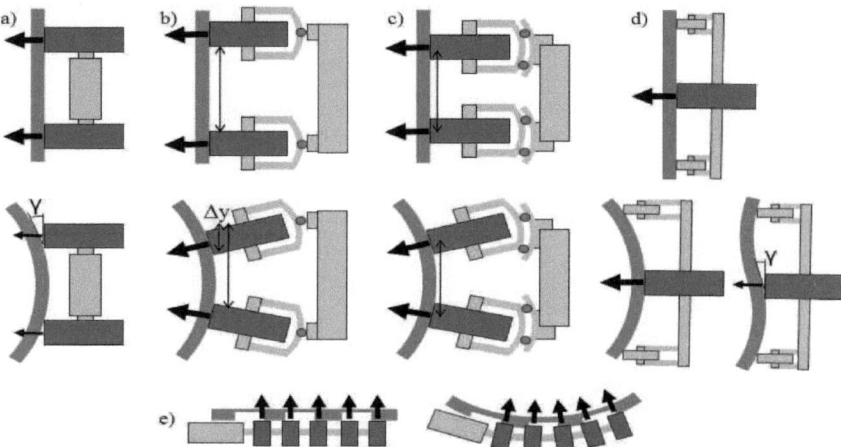

Fig. 4-27: Wheel suspensions for better adapting to surfaces with changing curvature (a) Adhesion force decrease in a simple magnetic wheeled vehicle without any wheel suspension, (b) Wheel suspension with only one joint per wheel, which leads to an unwanted lateral displacement (Δy) when turning on curved surfaces, (c) More complex mechanism with the suspension axes going through the contact points wheel/surface, (d) Active stabilization mechanism (MagneBike) and its force reduction on double-curved surfaces, (e) Several magnetic wheels on a flexible shaft, for moving in curved environments with small holes (e.g. generator stators)

For achieving a passive adaptation to surface while still keeping the possibility to span over small gaps with a vehicle of very small height, the newest generation of the generator air gap crawler uses flexible shafts with several magnetic wheels glued to it (see Fig. 4-27-e). By using a unit with several flexible magnetic shafts in serial (see Fig. 4-26-a-3), it is possible to roll over gaps in all directions and still adapt to changes of the curvature. Given the expected benefits in the highly valuable business of generator inspection (chapter 5.4.1), the technology of fixing magnetic wheels on flexible shaft is planned for a patent at ALSTOM [21].

4.2.3.2 Mechanisms for changing the direction of motion

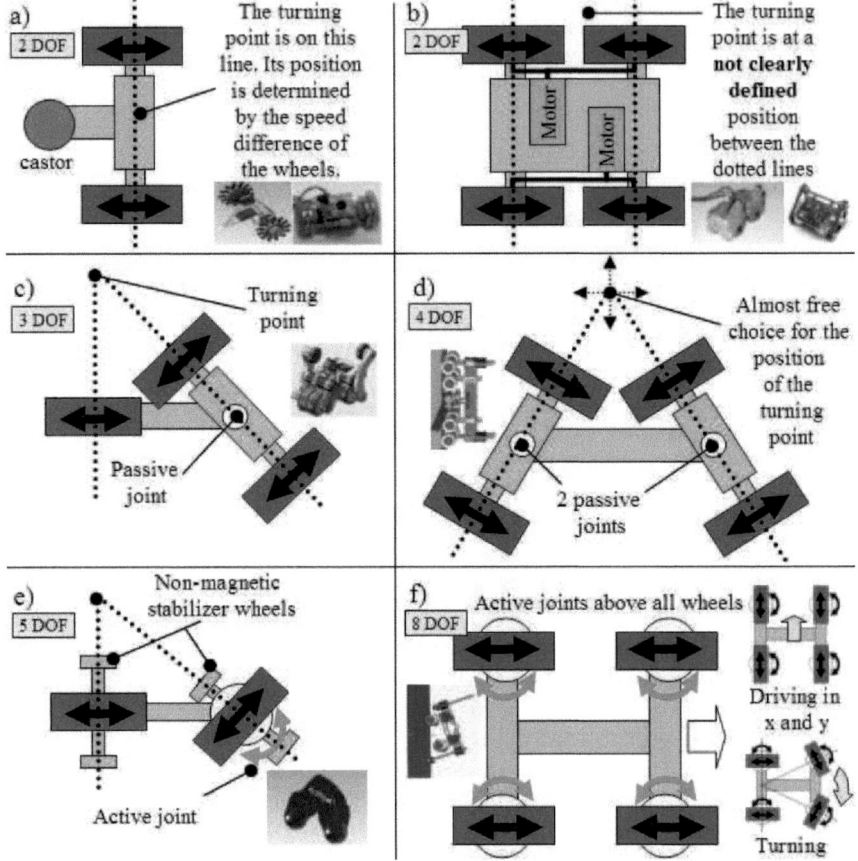

Fig. 4-28: The most common wheel configurations that are used in compact climbing robots, with pictures of representative prototypes
(a) Differential drive with two main wheels/whegs and a castor wheel/tail (Gecko [46], Micro-Tripod WpW [14])
(b) Differential-drive with several wheels on each side or caterpillars (First prototype for generator housings [12], TriPillar [39]),
(c, d) Three- and four-wheeled configurations with passive joints (First Prototype for the MagneBike PCAE [9], 2nd concept for gas tanks [2])
(e) MagneBike configuration with an active joint on the front wheel and active stabilizer wheels [7], (f) 4-wheeled configuration with active joints above all wheels (first concept for gas tanks [1])

For changing the moving direction of a compact climbing robot, the most obvious approach is to use classic steering configurations – similar to non-climbing robots. A selection of the most commonly used configurations in the field of compact climbing robots with magnetic adhesion is drawn in Fig. 4-28, while a more complete overview can be found in the chapter about "locomotion" in the book "Introduction to Mobile Robotics" [102].

Within these configurations, the simplest one with only 2 motorized DOF is to just use 2 motorized wheels or whegs, plus a castor. This configuration (see Fig. 4-28-a, usually called "Tripod") is the most frequently used in compact climbing robots with magnetic adhesion – mainly caused by its unbeatable simplicity. However, the mobility of such vehicles on transitions and combined obstacles (see pervious chapter) is limited, as the third contact point at the back (castor) cannot provide any traction force.

For increasing the performance on obstacles by using all-wheel traction, but still keeping only two motorized DOF on the vehicle, many robots use two units where several wheels are powered with the same motor (see Fig. 4-28-b). The torque-transmission is normally achieved with spur-gears (first prototype for generator housings [12]) or by caterpillars (TriPillar [39]). For steering, differential speeds are applied to each wheel unit – similar to the previously described vehicles. However, in this configuration the wheels (or caterpillars) slip relatively uncontrolled during the turning-operation, as an exact turning point is not defined by the vehicle geometry. For this reason, the error of localization by odometery is relatively high in vehicles of this type.

In order to achieve the advanced obstacle-passing-capability of all-wheel-traction and the relatively low slip during turning maneuvers, one approach is to power all wheels independently and add flexible joints in the vehicle structure – as drawn in Fig. 4-28-c and d. Examples for such configurations are the first version of the MagneBike PCAE [9] and the second concept for the gas tanks-scenario [2], both resulting in robots of relatively big size compared to others.

An approach for achieving a very compact design is the MagneBike-configuration (see Fig. 4-28-e and [7]), which only uses two motorized magnetic wheels and an active joint for turning one of them. As already explained in the last paragraph, for stabilizing this robot on traversing paths on vertical walls, two pairs of actively controlled extra wheels are used on each wheel (see Fig. 4-27-e). This robot structure can well adapt to very small curvatures. However, concerning lightweight design and achieving a low number of actuators it is not optimal – as only 2 of the 5 actuators contribute to the traction of the robot.

If there is not enough maneuvering space for turning the entire robot, another approach is to place active joints above each wheel for turning them separately (see Fig. 4-28-f). With such a combination, the robot is not only able to turn, but can also move in any direction without turning its main body. However, such robots require a large number of actuators and thus cannot be realized very compact. Examples are the first concept for gas tanks [1] and the PIR [48].

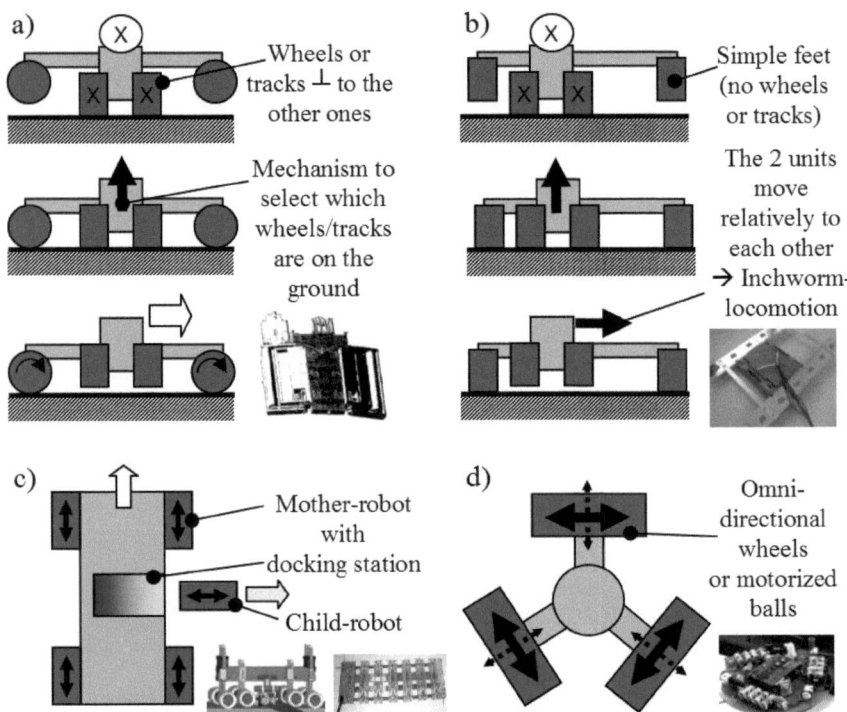

Fig. 4-29: Alternative concepts for changing the direction in compact climbing robots, with pictures of example prototypes
(a) Switching between two rolling units that point in different directions (concept by GE [59]), (b) Hybrid structure with magnetic wheels in one direction and inchworm-locomotion in the other one (Generator Air Gap Crawler with bi-directional mobility [18]), (c) Mother-child concept (applied for gas tanks [2] and the generator air gap [19]), (d) Robot running on three omni-directional wheels or balls (Tribolo [96])

4. Obstacle-passing with compact climbing robots

Besides the above mentioned standard configurations, also some alternative approaches for changing the moving direction have been implemented in compact climbing robots, with the most common ones drawn in Fig. 4-29.

For moving in two different directions without turning the robot, a concept that was developed by GE for the inspection of generator air gaps [59] uses a so-called "wing mechanism" for switching between two drive units – one for axial paths and one for circumferential paths (Fig. 4-29-a). A similar approach has also been followed by our team in the generator air gap crawler with bi-directional mobility [18]. Also this design uses a magnetic wheeled drive-unit for the axial direction and a mechanism for choosing which unit is in contact. In contrast to the GE-concept, this design uses simple feet instead of wheels in the second unit and achieves the motion in circumferential direction by relatively moving both units against each other in circumferential direction – which allows for an inch-worm-like gait (Fig. 4-29-b). With this design, it is possible to stay at very low height (our prototype: only 8mm) and to pass long gaps both in axial and in circumferential direction. However, the mechanical complexity of this concept is very high, and the motion in circumferential direction is not continuous.

Another approach for achieving motion in two directions by not turning the robot is the use of two robots in mother-child-configuration (Fig. 4-29-c). By avoiding the implementation of mechanisms for turning or switching between wheel units, both robots can be realized very simple (down to only 1 DOF), which finally results in a very low total complexity. Additionally, the child robot can be realized very fast and lightweight, which reduces the total inspection time and brings advantages in fragile environments. For this reason, the mother-child-concept has been proposed both for the projects on gas tanks [2] and generator air gap inspection [19] – providing the only feasible solutions at a time when the most advanced vehicle structures (MagneBike PCAE [9] and Air-Gap-Crawler double-flexible [21], both developed in 2009) were not available yet.

In non-climbing robot, a very frequently used alternative for turning the robot is the use of omnidirectional wheels (also called Swedish wheels), balls or other means of locomotion that provide traction in one direction and allow for free displacement in another (Fig. 4-29-d). With such a configuration, mobility in two directions plus turning can be achieved with only 3 motorized DOF (e.g. in the Tribolo [96]). To our best knowledge, in climbing robots only one prototype exists – the underwater-climbing-robot RIMINI [35].

4.2.3.3 Mechanisms for the entrance geometry

Another group of mechanisms has its application scope at the entrance to the component – as the most difficult obstacles sometimes occur only there. These obstacles can be avoided by using intelligent solutions for placing the robot already inside the part, which can allow for a vehicle design with much better performance at its final destination – leading to similar advantages as already explained in the context of the mother-child-concept. The mechanisms of this type that have been analyzed and designed for the robots on power plant inspection are represented in Fig. 4-30.

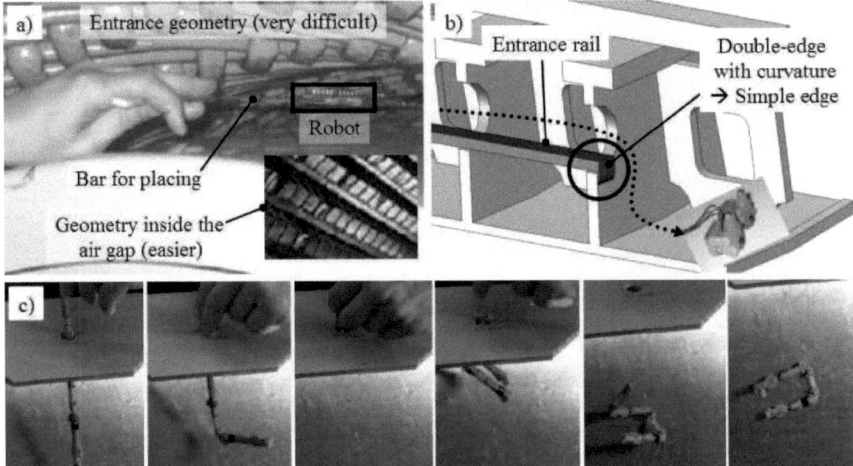

Fig. 4-30: Mechanisms for passing the challenges at the entrance
(a) Placing device in the air-gap-crawler [16], (b) Rail for avoiding the combined obstacles at the entrance gaps in the generator housing [12],
(c) Foldable micro-crawler for turbine inspection [24]

In both projects in the context of generator inspection, the difficult obstacles at the entrance are avoided by using intelligent placing devices – a bar that stays on the robot in the case of the air gap crawler [16], and a rail where the robot can roll on in the case of the back housing [12]. In turbine inspection, the main challenge at the entrance is the small size of the bore-scope holes (Ø15mm), which gets solved thanks to a foldable structure of the robot [24]. Technical details about these mechanisms are provided in the corresponding case-studies (chapter 5.3, 5.4.1 and 5.4.3).

4.3 Robotic vehicles and their performance

The mechanisms described in the previous section can be combined in different ways to form robotic vehicles. Starting with a brief classification that is based on the way the mechanisms are implemented in the robot, representative vehicle concepts from each main group are briefly described and compared in this section. Note that these descriptions only focus on the basic vehicle concept and the mobility on obstacles, while technical details about the prototypes and the industrial applications are discussed in the case studies (chapter 5). The section concludes with a brief comparison of approximately 30 representative prototypes based on typical obstacles that can be found in power plant components.

4.3.1 Classification based on the obstacle-passing-mechanisms

When analyzing the different obstacle-passing-mechanisms in more detail, it can be observed that these mechanisms are either placed around the contact area (wheel or foot) or in the robot structure. For designing a vehicle, mechanisms from both groups can be selected – and either in a passive or active way. Setting up a 3x3 matrix with the "mechanisms in the robot chassis" in the columns and the "mechanisms around the wheels" in the lines; and the three alternatives "nothing, passive and active" for both of them – the main conceptual design possibilities for climbing robots with obstacle-passing mechanisms can be represented in a very comprehensive way. Such an overview is shown in Fig. 4-31 - with pictures of the most important prototypes in each group.

What can be seen from this overview is the following: Most mechanisms that are implemented into the robot structure are mainly designed for passing edges or ridges, while most mechanisms around the contact area have the purpose to improve the performance on inner transitions (corners). Obviously, there is a tendency that not only the mobility but also the overall complexity (and robot size) gets increased when moving towards active solutions in both areas: Most of the very compact vehicle designs have either no mechanism at all (group a) or only a passive mechanism (group b) – which is not enough for passing the most difficult types of combined obstacles (thin ridges or sharp multiple steps). On the other hand, most robots with the ability to pass such obstacles need one or several active mechanisms for locomotion and result in relatively complex prototypes with large size (group c).

Fig. 4-31: Classification of compact climbing robots with high mobility, based on how and where the obstacle passing mechanisms are implemented, and with the most representative prototypes within each group (a) Robots without any special mechanism, (b) Robots with passive mechanisms, (c) Robots with active mechanisms, (d) Robots that combine active structures with passive mechanisms around the wheel for corners

Concepts that combine active and passive obstacle-passing-mechanisms in the same vehicle structure (group d) are not used very frequently yet – to our best knowledge only by our team: MagneBike PCAE and Micro-Tripod WpW.

Both designs use a passive mechanism for inner transitions (corners), and extensions with active joints in the structure for outer transitions and combined obstacles. With this combination, these robotic vehicles achieve impressive mobility at a relatively low system complexity. The basic idea of this combination can be seen as one of the main contributions in this work. Given the expected benefits in steam chest inspection, it is planned to register a patent which covers the most promising combinations within this concept [11].

4.3.2 Brief description of the most relevant vehicles

As already outlined in the previous section, the description of the most relevant vehicle concepts can be structured according to 4 main groups – simple robots without any special mechanism, robots using passive mechanisms, robots using active mechanisms and robots that combine active and passive mechanisms:

4.3.2.1 Simple robots without special mechanisms

As already explained in chapter 3 – many climbing robots used for inspection purposes do not have any mechanism for obstacle-passing implemented. If the surfaces are almost plain (e.g. oil storage tanks) or always with the same curvature (e.g. turbine rotors), either magnetic wheels, magnets in the chassis (e.g. MRS 100 [33]) or even alternative adhesion principles like sliding vacuum suction (e.g. City Climber [82]) can be used for the adhesion of a wheeled robot. However, when dealing with slightly more complex environments with smaller curvatures, magnetic wheeled vehicles with suspended wheels are normally the principle of choice (e.g. the MRS 200 [32]). Simulations and tests with magnetic wheeled vehicles using all-wheel-traction and a high-friction rubber-cover on the wheels showed, that such simple vehicles are even able to pass 90° inner transitions without extra mechanisms – however only on clean surfaces where a friction coefficient of $\mu > 0.8$ can be assured. More details about such simple vehicles can be found in our paper about these simulations and tests [3] and the state-of-the-art-overview in the case-study about steam chests (5.2.3.1).

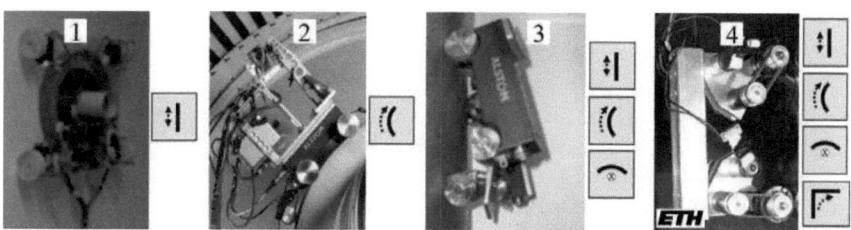

Fig. 4-32: Robots without any special mechanism
(1) City-Climber with pneumatic adhesion in the chassis [82], (2) MRS 100 with magnets in the chassis [33], (3) MRS 200 with magnetic wheels [32], (4) First test-prototype with magnetic wheels and all-wheel traction [3]

4.3.2.2 Robots that use passive mechanisms

For passing such inner transitions also on wet/dirty surfaces or by using wheels without rubber, several mechanisms have been developed and implemented into robots. While the detailed description and comparison of all these mechanisms (also against active mechanisms) can be found in the previous section about obstacle-passing-mechanisms (4.2.1), the photos of their corresponding robot prototypes are shown in Fig. 4-33, with the references in the figure caption.

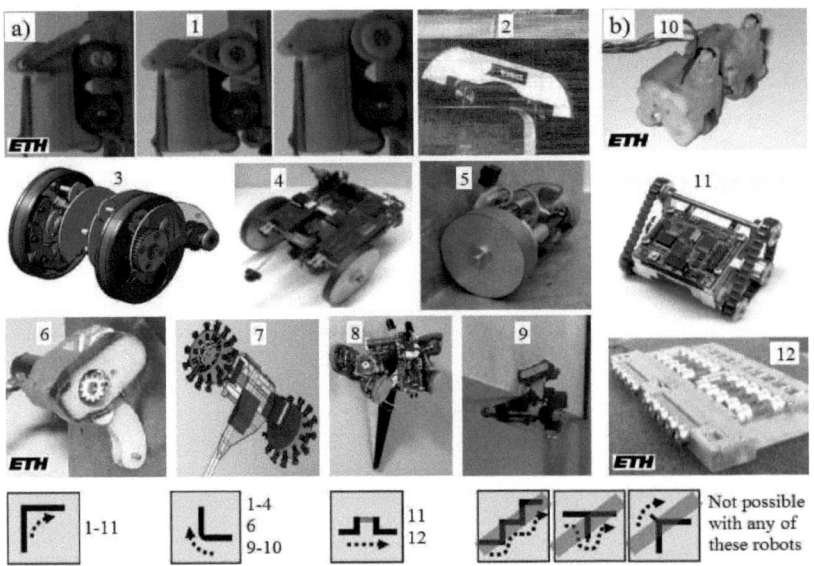

Fig. 4-33: Robots with passive mechanisms
(a) around the wheels or (b) in the robot structure
(1-11) Mechanisms for corners (11 and 12) Mechanisms for holes
(1) Modular test prototype for passive front arm, Tri-Wedge, WpW [10],
(2) Sewage pipe robot with dual magnetic wheel [40], (3) First student prototype at EPFL using the multiple wheel-inside-the-wheel-concept [42],
(4-5) Robots with a passively moving adhesion zone for corners [44,45],
(6) Rotating magnetic cam disc [15], (7) "Gecko" with special magnetic wheels [46], (8) WaalBot running on whegs with artificial gecko-hair-pads [84], (9) Gel-type sticky mobile inspector [85], (10) First ETH-design for generator housings with passive extra wheels in the structure for passing corners [12], (11) TriPillar with magnetic tracks instead of wheels for passing small holes in the surface and a triangular shape of these tracks for better passing corners [39], (12) Generator Air Gap Crawler with several wheels in a row for passing small holes [12]

Note that also the use of caterpillars instead of wheels (or several wheels in serial) – with the goal to passively roll over surfaces with gaps (see Fig. 4-26-1) – is a passive mechanism for obstacle passing. For this reason, robots of this type (e.g. the first Generator Air Gap crawler [12]) are also listed in this context.

What can be observed when analyzing these robot designs is the following:

- Almost all of these robots (exception: robot 2, for sewage pipes) are still "first-generation-prototypes" with the main purpose to prove the functionality of the mechanism and to show what minimum robot size can be achieved with it.

- For this reason, they are mainly built in a very basic configuration – only a propulsion unit with the mechanism, a second unit that is symmetrical to the first one (for turning the robot), and in some cases also a battery, electronics for remote-control and/or a camera.

- This basic design allows for sizes in the range of around 30mm-50mm, which is significantly smaller than it can be achieved with the most compact existing robots that use active principles (e.g. the MagneBike, 200mm) – making environments accessible that are too narrow for other robot types.

- On the other hand, the mobility is also not as high as in robots that use active mechanisms: Complex combinations of steps, ridges or even sharp edges with non-magnetic zones cannot be passed. Many robots even fail on normal edges.

4.3.2.3 Robots that use active mechanisms

For achieving a higher mobility on these types of obstacles, the most logic way is to use active mechanisms instead of passive ones. Unlike in rough terrain locomotion – where active vehicle structures often lead to a complexity that is not reasonable for industrial use any more (e.g. in the Octopus [89]) – in magnetic wheeled climbing robots the additional complexity that comes along with active vehicle structures still remains reasonable. This can be seen in the following examples (robot pictures in Fig. 4-34-a/b):

- By adding just one additional DOF to the wheel units in the MagneBike [7], this robot is able to pass corners and difficult combinations of steps with no extra restrictions concerning the friction coefficient between wheel and surface.

- With just two additional linear actuators, the capability to pass thin ridges with saturation effects (one of the most difficult obstacles in power plant environments) could be achieved in the first concept of the gas-tank-robot [1].

- In the Generator-Air-Gap-crawler with bi-directional-mobility [18], the extension with inchworm locomotion in circumferential direction only brought the addition of two extra actuators and allowed for still keeping the total vehicle height below 9mm.

Even when using both an active mechanism for force reduction AND active joints in the structure, the control complexity still remains reasonable compared to similar systems in non-climbing robots (e.g. humanoid robots) – mainly for the reason that in climbing robots with strong adhesion principles and relatively slow motion, unfavorable dynamic effects on the control are almost irrelevant. For this reason, several hybrid legged/wheeled robots and also pure biped robots have been realized – mainly for applications where size, mass and speed are not very important. On specific types of obstacles, these robots usually achieve higher mobility than wheeled climbing robots (see Fig. 4-34-c).

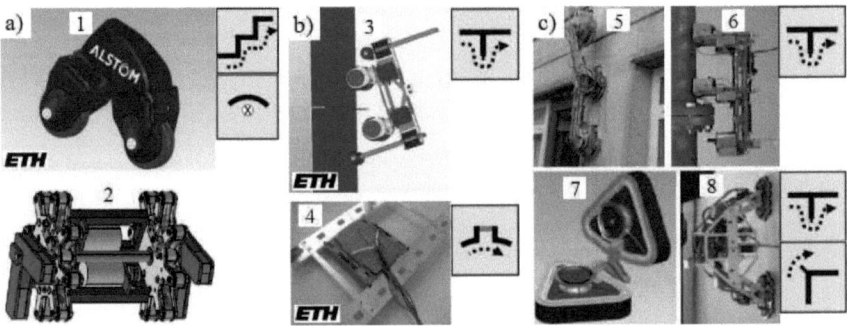

Fig. 4-34: Robots with active mechanisms (a) around the wheel, (b) in the robot structure, or (c) both around the wheel AND in the robot structure (1) MagneBike [9] with active rotary lifters that are also used for stabilizing, (2) SpokeHeel [43] with a moving adhesion zone for corners and active actuation, (3) Robot for gas tank inspection with 2 active linear actuators in the structure for passing ridges [1], (4) Generator Air Gap Crawler with extension for bi-directional mobility using inchworm-locomotion for the circumferential drive [18], (5-7) Wheeled robots with active joints in the structures and mechanisms for active force reduction – Alicia³ [80], PIR [48] and City-Climber [82], (8) Roma II [75] in biped configuration

However, the most severe disadvantage of active mechanisms in climbing robots should be mentioned again as well – the high mass of the additional actuators. This relatively high mass comes from the need that these actuators usually have to provide high forces in the range of the adhesion force (usually 5-10 times

4. Obstacle-passing with compact climbing robots   103

higher than m*g, for assuring enough security against slipping). Sometimes these forces have to be actuated at positions that are not easy to access – for example in the MagneBike [9], where lifter-axis and wheel-axis need to be coaxial.

More details about such specific design challenges can be found in the corresponding case-studies (chapter 5), but a preliminary conclusion can already be drawn here:

> The increased mobility when using active solution usually comes with a significantly increased need for mass and space – which makes it very difficult to realize such vehicles at sizes below 100mm in the largest direction.

4.3.2.4 Robots that combine active and passive mechanisms

Given the advantages of both the active mechanisms (high mobility) and the passive mechanisms (small size and mass), it seems straightforward to combine both of them for designing compact climbing robots with outstanding performance. This approach is still very new and to our best knowledge only realized in the two most recent prototypes that have been developed in our team – the Micro-Tripod WpW [14] and the MagneBike PCAE (PassiveCorner/ActiveEdge) [9].

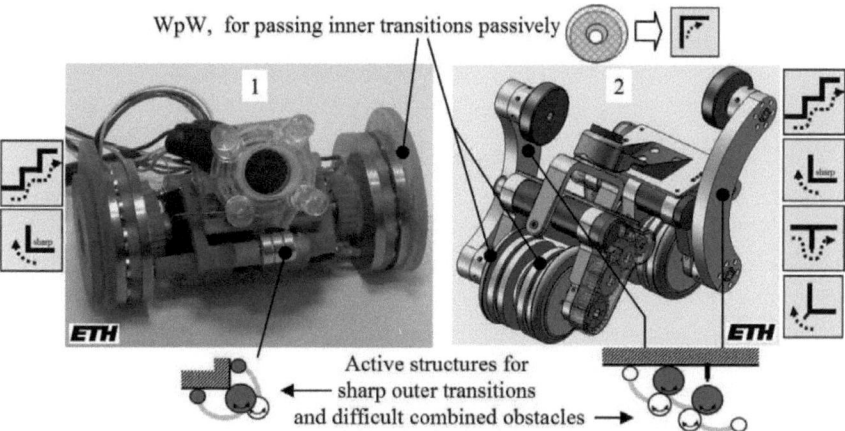

Fig. 4-35: Robots that combine active structures for passing edges with passive mechanisms around the wheel for corners (1) MagneBike PCAE with new wheel units that use the WpW technology and an active extension-arm for edges and ridges [9], (2) Micro-Tripod WpW with its active edge-extension-arm coupled with the camera movement [14]

Both robots use the recently developed WpW-mechanism [10] for passively rolling through corners, while active joints in the structure are used for increasing the normal force of the main wheels when passing sharp edges (outer transitions) – which increases both the mobility and the payload capability. The MagneBike PCAE is even able to pass ridges and edges with a short non-magnetic zone – obstacles that had previously been impossible for robots at such small size and with this low mechanical complexity (< 200mm long, 5DOF).

While the MagneBike PCAE has been mainly realized as an upgrade for the original MagneBike [9], the Micro-Tripod WpW is the latest generation of the robots that were designed for accessing the back housing of generators [12-15]. Details about the design of both robots can be found in the corresponding case-studies (chapter 5.2 and 5.3).

4.3.3 Comparison

Even if the two most recent robots that combine the WpW-technology with active structures are already very outstanding regarding their high mobility at very small size, they are still not able to deal with all size restrictions and obstacles that can be found or even imagined in power plant components and similar applications. In specific environments such as generator air gaps or boilers, other prototypes that are specialized for these applications still perform better. Future developments can once also show similar or even better performance than these two robot concepts. For this reason, a tool for comparing the performance of compact climbing robots has been developed as well.

This comparison tool is based on a matrix, with the applications and prototypes labeling the columns, and the most relevant environment challenges labeling the lines. For all environment challenges, both the significance in the industrial applications and the performance of each robot is entered at the corresponding cell of the matrix – with the robot performance mainly based on experiments with the prototypes, in the case of external developments sometimes also on calculation models, video analysis or discussions with the developers. Additionally, a color code is used for linking the "significance" and the "performance"-information for providing a better overlook at first glance.

With this data, both the difficulty of each challenge and the relative performance of each robot can be seen in a complete but still comprehensive way. A commented excerpt of this matrix is drawn in Fig. 4-36.

4. Obstacle-passing with compact climbing robots 105

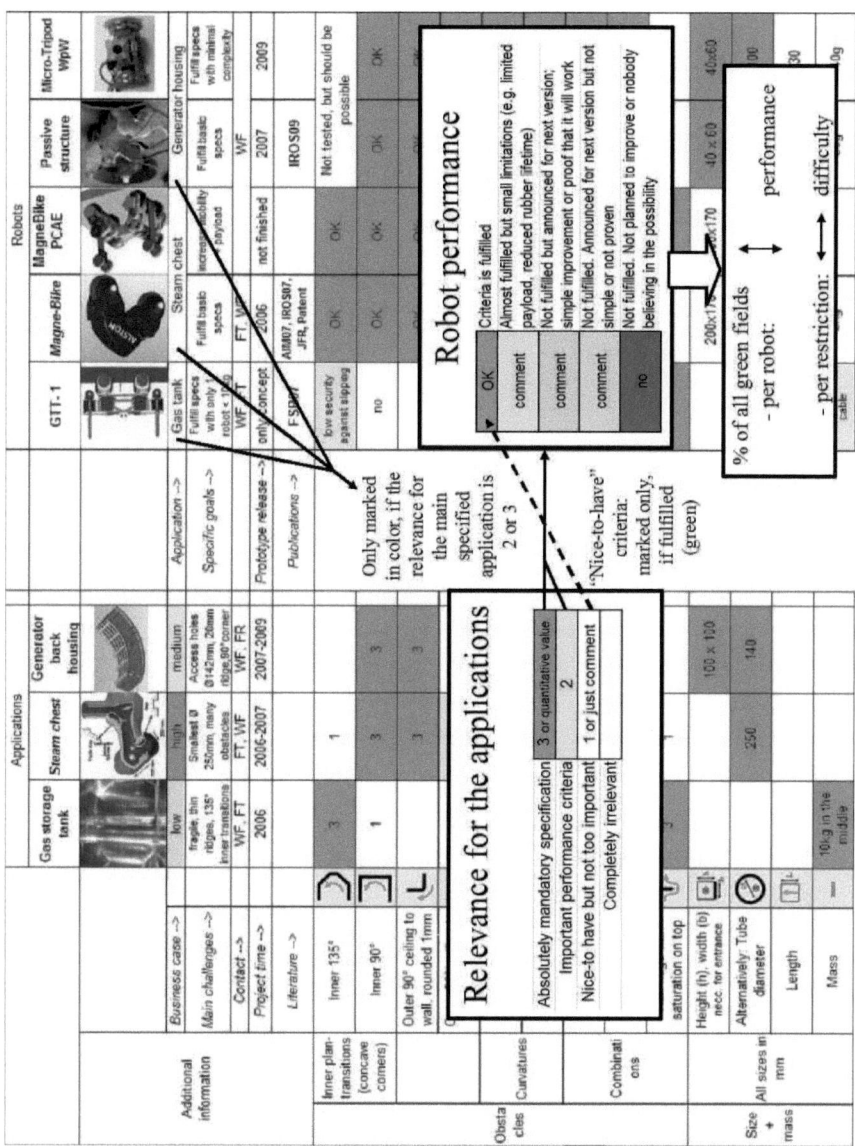

Fig. 4-36: Excerpt of the matrix, with comments how the fields "relevance for the application" and "robot performance" are filled; and how the overall performance of a robot and the difficulty of a specific challenge can be estimated

At the current state of the project, the comparison matrix contains 7 different types of applications that lead to 43 different types of environment challenges. Up to now, 30 robot prototypes are evaluated according to these criteria. The full matrix at readable size (A2; split into 4 A4-pages) can be found in the appendix.

4.4 Summary and conclusion

In this chapter, the most typical obstacles and hazards for climbing robots in power plant components have been analyzed and classified – based both on the requirements coming from real applications, and on the performance of the most recent robot prototypes. The most important obstacle-passing-mechanisms and vehicle structures are described and classified as well; and compared according to their performance in the above mentioned environment challenges. For facilitating the comparison, a tool based on a matrix-representation has been established.

What can be concluded from this chapter is the following:

- Given the high number of obstacle-passing-mechanisms (already 18 only for corner passing) and relevant robot prototypes (30 analyzed in this work), the structured classification of all this work and the objective comparison based on the environment challenges in real applications will become an important help for future research teams that have to design innovative climbing robots in the field of power plant inspection and similar applications.

- The two most recent robots which combine the WpW-technology with active structures [9, 14], show significant advantages over previous designs regarding their high mobility at relatively small size and complexity.

- However, they are still not able to deal with all environment challenges that can be found in power plant components and similar applications. For this reason, the comparison matrix also includes the special developments for such types of applications (e.g. boiler tubes or generator air gaps [16-21]).

While this chapter mainly stressed on the conceptual description of all mechanisms and prototypes, the detailed technical description of specific design challenges (e.g. torque-transmission at very small size) plus the link between real business cases, detailed environment specifications and specific robot prototypes is addressed in the case-studies (next chapter).

5 Case studies

The following chapter provides an overview on the projects on compact inspection robots that have been realized at ASL from 2005 to 2009 – with the three technically most interesting ones (see Fig. 5-1) described more detailed.

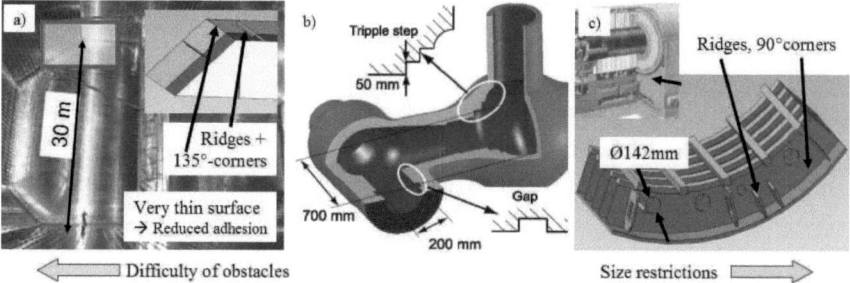

Fig. 5-1: The three scenarios with the most difficult challenges for robot mobility (a) Gas storage tank, (b) Steam chest, (c) Generator housing

As already described in 2.1.1, all of these scenarios need a robot mobility that allows for passing relatively difficult obstacles, while the allowed maximum robot size is highly restricted by the small available space. For this reason, new obstacle-passing-mechanisms, vehicle structures and robot prototypes had to be developed – mainly the ones described in chapter 4.

The first case study is about the gas-tank-scenario, where the encountered obstacles are the most difficult but the robot size is not restricted. It is followed by the steam-chest-application, where the obstacles are less difficult, but the size is restricted to only 250mm. The third case study is about the most challenging environment of all – generator housings – where the allowed size is even smaller than in steam-chests, but the obstacles almost as difficult.

All these three case-studies are structured as follows: At the beginning, the business-case and the previous inspection methods are described; as well as the most important specifications and challenges for the robot's locomotion. This specification analysis is followed by a description of the state of the art before the project start and the different prototype generations, focusing on the core advantages towards previous designs and how the most difficult design challenges have been solved. Each case study concludes with a summary of the core innovations and some "lessons-learned" for future projects.

In addition to these three main case studies, the last subchapter provides a brief overlook on other projects in the field of power plant inspection where successful prototypes have been realized in the context of this thesis – generator air gap inspection, boiler inspection by using micro-helicopters with docking-capability, micro-robots for tube- and turbine inspection and power-line inspection.

5.1 Gas storage tanks

The project on gas tanks was the first project on magnetic wheeled robots (2005). It was funded by Gaz-Transport and TechniGAZ (not ALSTOM Power, as all others) and also the environment is not directly located in a power plant. However, the challenges are highly similar to the "real" power plant applications that are described in the next sections. Additionally, many innovations in the other case-studies are derived from basic conceptual ideas in this project.

5.1.1 Motivation and business relevance

Gas tanks made out of thin sheet metal are installed in oversea ships that are used for the transportation of liquid gas. Periodically, these tanks have to be inspected for detecting leaks, especially along the welds. For this purpose, helium is injected in the structure that surrounds the tanks. A sensor that is able to detect helium leaking into the tank is used to determine the position of leaks. Currently, this sensor is carried by a balloon that is operated manually with the use of long ropes. As this method is very slow and imprecise, a better inspection system had to be developed, preferably using wall climbing robots with magnetic wheels.

5.1.2 Environment specifications

The goal is to develop an inspection system that can carry a sensor module (approximate mass: 500g) to every point of the interior hull of a gas tank (Fig. 5-2). The robot must therefore be able to move vertically as well as upside down, suspended from the ceiling. It must be able to pass obstacles in the form of ridges and 135°-transitions and must be built light enough to not destroy the fragile structure of the tank.

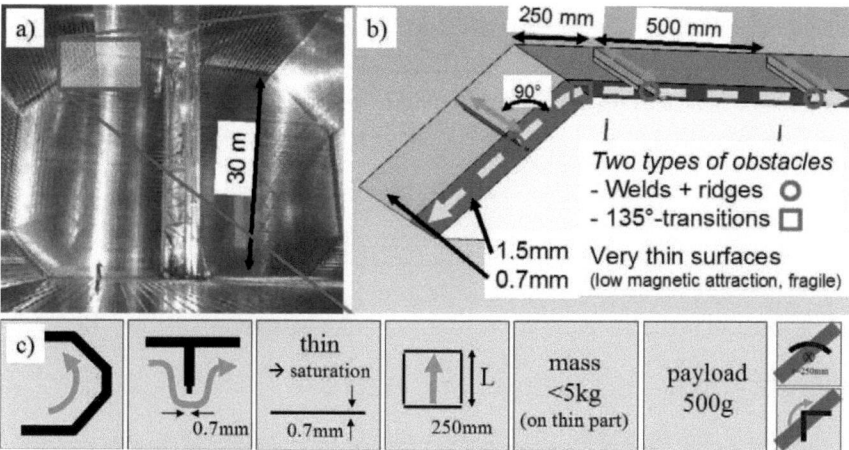

Fig. 5-2: The environment to inspect, with the most difficult challenges –
ridges, 135°-transitions and very thin surfaces with low magnetic adhesion
(a) Photo, (b) CAD-image, (c) The most difficult environment challenges

5.1.2.1 Obstacles (ridges and 135°-inner transitions)

The main obstacles that the robot has to pass are ridges and 135°-transitions (see Fig. 5-2-b, dashed line). On top of the ridges (the worst obstacles) the magnetic attraction gets reduced to approximately 10% of the normal value (also see Fig. 4-4). As already explained in the chapter about vehicle structures (4.2.2) and in our first paper at FSR07 [1], a structure with more than 2 pairs of wheels and active joints in the chassis is necessary to pass them with a wheeled robot.

5.1.2.2 Fragile structure in the center

Normally, the central area of these tanks (see Fig. 5-2-b, light area) consists of very thin (0.7mm instead of 1.5mm) sheet metal, which is only attached at the ridges. Thus, a huge and heavy robot would plastically deform and damage the structure. Detailed FEM simulations and tests in the environment (performed by our industry partners) showed that the thin 0.7mm-sheets can only support a mass of 5 kg, while the thicker ones (1.5mm) at the borders to the next surface can support a robot mass up to 20kg.

5.1.2.3 Need for a relatively high speed

In order to keep the total inspection time low, another important goal was to achieve a speed of approximately 10 m/min – at least in the horizontal direction parallel to the ridges.

5.1.3 Prototypes

In the context of the core project (2005/06), three groups of concepts have been developed and partially implemented in prototypes. After the project, also the MagneBike PCAE was tested successfully in this environment (2010).

5.1.3.1 Related prototypes and state of the art in 2005

For getting the project started, the assumption was stated that the obstacle-passing-mechanisms in rough-terrain rovers (e.g. the Octopus [89], Fig. 5-3-a) and magnetic wheeled climbing robots could be highly similar. As at ASL there was lots of experience in the field of rough-terrain rovers, it seemed straightforward to propose magnetic wheels on one of these structures for convincing the responsible project managers.

Driven by this assumption, a preliminary prototype with passive suspension and magnetic wheels was realized by other researchers at EPFL – mainly reusing parts from old prototypes (Fig. 5-3-b). With a suboptimal wheel design of very low adhesion force, the motors under-dimensioned around factor 5 and no suitable obstacle-passing-mechanism implemented, its performance was still far away from the real specifications: The robot could not even climb vertical walls, but only roll over small bumps or ridges on horizontal surfaces, while the mass was already 3kg. Improving this robot in order to fulfill all specifications was regarded as a relatively simple task and assumed to approximately 2 months of work.

When looking at the state of the art in magnetic wheeled climbing robots in the year 2005, some prototypes were already industrialized and commercially available. Almost all these robots (e.g. he Tripod by Jireh Industries [31], Fig. 5-3-c) use magnetic wheels in the standard design; with an axially polarized ring magnet and two steel rims (as described in [28] and Fig. 3-5). However, they do not have any obstacles passing mechanism implemented and thus cannot pass ridges, sometimes not even 135°-transitions.

5. Case studies

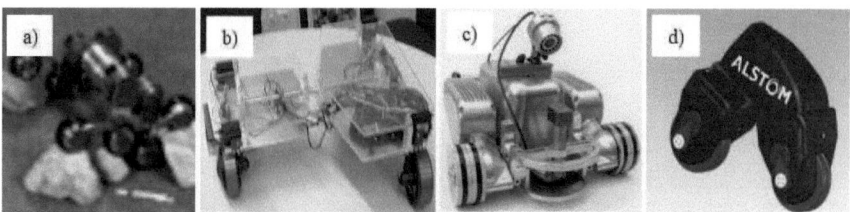

Fig. 5-3: Related prototypes analyzed at the beginning of the project on gas-tank inspection
(a) Octopus [89], (b) Preliminary prototype for first analysis, (c) Tripod by Jireh industries [31], (d) MagneBike [7]

Parallel to the gas-tank-project, also the first project for ALSTOM on the steam-chest-scenario (chapter 5.2) had started at ASL. Even if the first concept for this scenario (the MagneBike [7], Fig. 5-3-d) can be seen as an important milestone in the field of compact magnetic wheeled robots, its mobility is still not high enough to pass the worst-case obstacle in this scenario – thin ridges.

The only prototype that already existed in 2005 and that had the ability for dealing with this type of obstacle is the "Pipe Inspection Robot (PIR)" [48] (see Fig. 5-4-a). It uses 3 pairs of wheels that can be lifted with linear actuators for passing steps, ridges or other combined obstacles on the outside surfaces of pipes. For choosing the right wheel to lift, each wheel unit is equipped with an additional mechanism for active force reduction (also called "jack-mechanism", more details see 4.2.1.1). The mechanical complexity of this robot is quite high – resulting in already 9 active DOF just for passing the ridges – without any mechanism for steering implemented yet. Its mass is 5.26kg – without inspection sensor and control electronics, and wheels that are not optimized for the thin surfaces with saturation problems yet.

5.1.3.2 First design with simplified structure for passing ridges

For realizing almost the same mobility, but at lower mass and with fewer active DOF, we developed a vehicle structure that does not need a mechanism for active force reduction any more (see Fig. 5-4-b). For passing the ridges (obstacle-passing-sequence, see Fig. 5-4-c), it only needs two linear actuators and the motorization of half of its wheels – which sums up to only 4 active DOF instead of 9 (PIR). For moving in any direction without needing too much maneuvering space, all wheels can turn separately.

112 5.1. Gas storage tanks

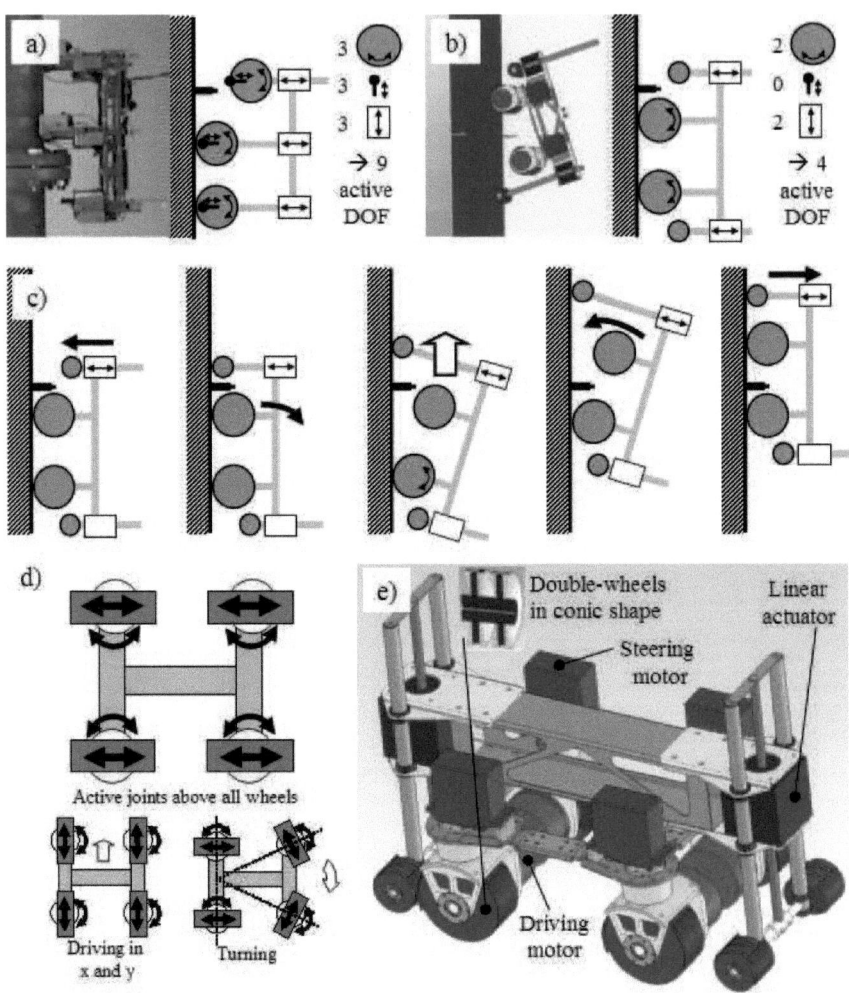

Fig. 5-4: The first design for the gas-tank-scenario [1], its basic mechanical concept and advantages towards previous systems,
(a) Previous design with significantly higher complexity (PIR [48]),
(b) Vehicle structure with fewer active DOF for passing ridges,
(c) Motion-sequence when passing a ridge-type obstacle,
(d) Steering concept that allows for driving in any direction without the need for maneuvering space,
(e) CAD-model of the structure, with the core components highlighted

5. Case studies

With all parts mass-optimized and the wheels adapted to the very thin surfaces (thin double wheels in conic shape, see Fig. 3-5-d), the expected mass of this concept was estimated to 10kg – still too heavy for the fragile environment. Reaching the specified speed of 10m/min would not have been possible either with the high reduction gears that are necessary for vertical climbing.

5.1.3.3 Mother child-concept

As pointed out in the previous paragraph, the use of a single robot that is able to go everywhere in the tank did not look very promising – especially when considering the results coming from a more detailed FEM-analysis that had been done after the presentation of the first concept: Only 5kg instead of 10kg were allowed for the robot mass when cruising on the very thin sheet metal in the middle of the tank. Also the slow speed of the first prototype was a problem.

These disadvantages led to the idea of separating the inspection system into two robots with specialized tasks. The smaller of these two robots is built very simple and without the ability to climb. It just moves along horizontal paths and uses the ridges in the structure as a guidance rail. Because of its low mass it can roll on the fragile surfaces without causing problems. For passing to the next ridge, it docks at the bigger robot. This robot always stays on the thick sheets (1.5mm) near the rims. Its mass is less critical and the robot can hence be realized complex enough for being able to climb vertical and pass difficult obstacles such as thin ridges.

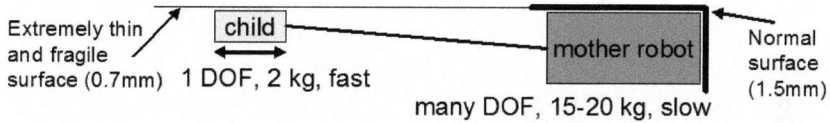

Fig. 5-5: Basic idea of the mother child concept and its application in the gas-tank-scenario

Another advantage of this approach is the possibility to realize the child robot very fast. As the robot is light-weight and only moves horizontally, the required torque is very low. Thus, a fast actuator with low reduction gear can be chosen.

For docking and undocking the child robot to the mother, the same linear actuators can be used as for passing the ridges. As the child robot is released perpendicular to the moving direction of the mother robot, also turning on spot can be avoided – which again decreases the mechanical complexity of the overall system. More information about the detailed design of the two robots, the control concept and some preliminary tests of subsystems can be found in our paper at IROS08 [2]. The sequence of releasing the child robot can be seen in Fig. 5-6.

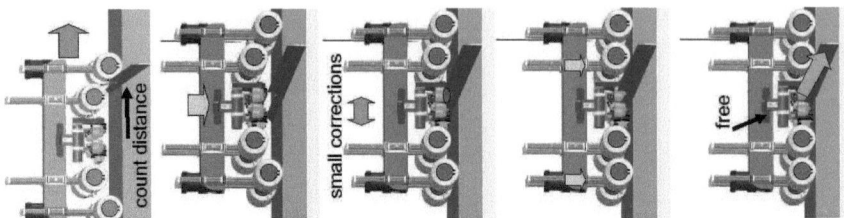

Fig. 5-6: The second design for the gas-tank-application, with two robots in mother-child-configuration [2] – sequence of placing the child robot

5.1.3.4 Industrialization of the child robot

For the last steps towards industrialization, the project was split into two phases – design of the child robot, and design of the entire system. The first phase was done in collaboration with BlueBotics and finished successfully, with a robust industrial prototype at the end.

The second phase was not completed any more, as it turned out that placing the child robot manually by a worker who climbs up on a ladder was already sufficient for satisfying the industrial needs.

5.1.3.5 Tests with the MagneBike PCAE in this environment

More than two years after the official project end, a robot with the originally required mobility and less than 5kg of mass was available – the MagneBike PCAE [9] which is the latest generation of the prototypes in the steam-chest scenario.

Preliminary tests and calculations showed that its outstanding mobility does not only bring significant advantages in the steam chest environment, but also allows for passing the required obstacles in the gas-tank application.

As the MagneBike PCAE was mainly developed for the steam-chest-scenario, the technical details about this robot can be found in that case study (see 5.2.3.4).

5.1.4 Innovation, outlook and lessons learned

5.1.4.1 Innovations in the core project

In the core project from 2005-2006, mainly two conceptual ideas can be regarded as highly innovative: The active vehicle structure for passing ridges [1] and the mother-child concept [2]. Both approaches allow for significantly reducing the mechanical complexity of a robotic system while keeping the same mobility as in previous solutions. The mother-child concept additionally allows for a significant decrease of the mass in the fragile zone of the environment and an increase of the robot speed there. Even if both concepts have not been industrialized at the end, their basic ideas could be promising in other applications.

5.1.4.2 Fundamental ideas that lead to innovations in other projects

Many of the central innovations in the following projects have their origin in basic ideas that had been created during this project – with many of them leading to patent applications and successful prototypes. The following list provides a short overlook on all three of them and points to their economic relevance:

- The basic vehicle structure for the MagneBike PCAE [9] was derived from the vehicle structure used in the first concept for the gas-tank-robot. Integrating this new structure into the MagneBike-prototypes is relatively low effort, but allows for a significantly increased mobility.

- Discussing the mother-child-concept also for the project on generator air gaps finally led to the innovation of the flexible magnetic shaft [21] that was mainly inspired by the design of the child robot in this application. With this technology, the newest generation of the air-gap-crawler could get simplified significantly – which finally allows for a design that is robust enough for real industrial use.

- Also the idea for the rotary lifter [8] had originally been generated in the context of the gas-tank project. It allowed for realizing the MagneBike, which was the first robot that achieved the required mobility in steam chests and similar environments of high economic value for robotic inspection.

5.1.4.3 Industrialization and outlook

A simplified version of the original design has been successfully industrialized by BlueBotics. Even if this design does not contain the innovative components from the original version any more, it well satisfied the customer. In the future, the industrialized version of the MagneBike PCAE could be used in this application as well – for finally achieving the originally specified mobility.

5.1.4.4 Conclusion and lessons learned

The most important lesson that can be learned from this project is the finding that there is a significant difference between the obstacle-passing-mechanisms in rough-terrain rovers and climbing robots on magnetic wheels.

Other important experiences coming from this project are the observation that all specifications should be re-asked critically during the entire project, and that innovation can sometimes go beyond pure vehicle design: Only by questioning the almost self-evident but relatively unnecessary goal of just using one robot in the system; and through re-asking the allowed mass in more detail (different values for the two types of surfaces), the new and advantageous concept of the "mother-child-structure" could be developed.

What can additionally be concluded is the recommendation that innovative ideas from previous projects should be well documented. By reusing them in the following projects, impressive results could be achieved with relatively little effort there (see the other case studies and Fig. 5-36).

5.2 Steam chests

5.2.1 Motivation and business relevance

The steam chest is the component that distributes the hot steam coming from the boiler before entering a steam turbine (Fig. 5-7-a). Given the combination of high temperatures, pressure, and vibrations coming from the turbine, these parts are highly stressed and thus need to be inspected regularly for small cracks or other failures. In contrast to the steam turbine itself, steam chests are made out of just one part and thus cannot just be opened for the inspection but have to be accessed through the pipes coming from the boiler. For inspecting these parts, stiff bore-scopes cannot be used because of the high number of bends and intersections, and flexible ones either because of the large diameter of the pipes. Standard in-pipe inspection robots that spread from one side to the other cannot be used either, as most of them cannot deal with the large and abrupt diameter changes. For these reasons and in order to exploit promising technology, this project very soon also focused on the development of magnetic wheeled robots with high mobility.

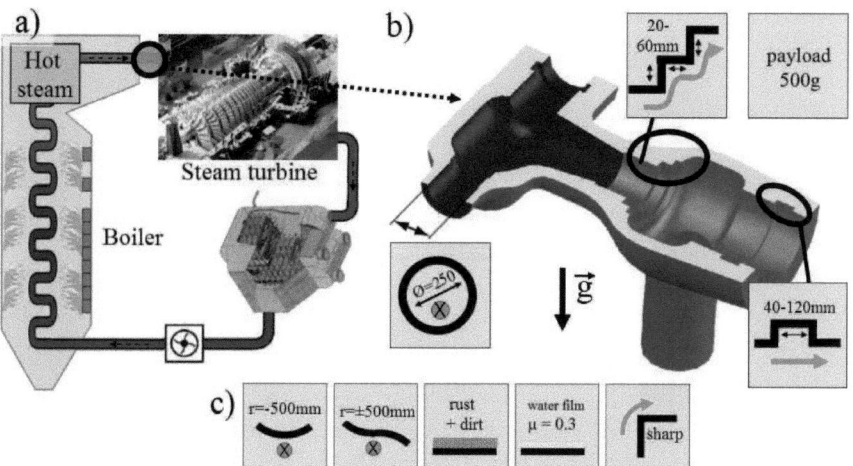

Fig. 5-7: The steam-chest application
(a) Position in the power plant, (b) CAD model showing the most important specifications - difficult obstacles such as tripple steps and gaps, the smallest size restrictions and the payload
(c) Non-mandatory options for extending the application scope

5.2.2 Environment specifications

For defining the environment specifications, a CAD model was derived from old hand-drawings and analyzed in terms of size restrictions and typical obstacles (Fig. 5-7-b). While a maximum mass was not specified for this environment, only the payload was defined to 500g. Concerning additional hazards, also wet and rusty surfaces were specified later in the project, as these challenges were found in the test environment for the first prototype. After these tests, also additional obstacles such as concave curvatures and sharp edges have been added to the specifications list for the second generation (MagneBike PCAE [9]), as the first prototype (MagneBike [5-8]) showed some limitations there.

5.2.2.1 Combinations of inner and outer 90° transitions

The typical "worst-case" obstacles that can be encountered in steam chests are mainly formed by abrupt diameter changes, which can be seen as combinations of concave corners and convex edges – with the typical step height being in the range of 20-60mm. More complex combinations such as tripple steps or gaps can be encountered as well. Note that most of these obstacles do not only occur between two plain surfaces but in curved pipe environments with diameters down to the minimum of Ø250mm and in all inclinations in respect to gravity – making them even more difficult to pass.

Sometimes, also rust, dirt or water can be encountered – with the two negative effects of reducing the lifetime of the rubber (especially rust) and of decreasing the friction coefficient between wheel and surface from $\mu \approx 0.8$ to $\mu \approx 0.3$. In some steam chests, the outer 90° transitions can also be very sharp – resulting in a significant reduction of the magnetic adhesion force there (down to 25%).

5.2.2.2 Size restriction and payload

The pipe diameter normally varies between Ø250mm and Ø800mm. As already described in the last section, the high ratio between biggest and smallest diameter is mainly limiting for robots that mechanically spread – which justifies the use of magnetism for the adhesion to the wall. The allowed maximum robot size is determined by the value of Ø250mm for the smallest diameter.

For being able to carry most types of NDT-sensors and some extra sensors for improving the localization capability of the robot (e.g. a laser range finder), the payload capability was specified to 500g.

5. Case studies   119

5.2.2.3 Curvatures

As the environment mainly consists of pipes, the surface is usually not flat but curved – with the curvature mainly being concave between flat and the minimum pipe diameter (Ø250mm). In some sections of the steam chest, the robot also has to drive on convex curved surfaces or combinations of convex and concave.

5.2.3 Prototypes

In total, three prototypes have been developed for the steam-chest-scenario – which can be seen in Fig. 5-8 – a simple test prototype for first analysis [3], the MagneBike [5-8] and the MagneBike PCAE [9].

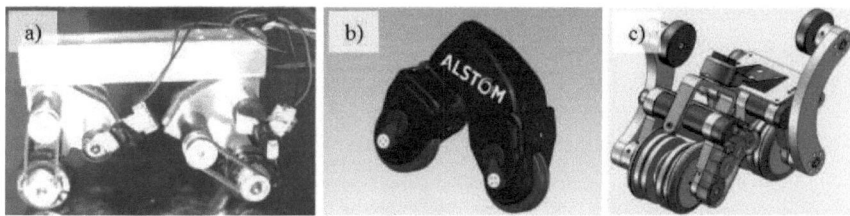

**Fig. 5-8: Prototypes developed for the steam-chest-scenario
(a) Simple test prototype for first analysis [3], (b) MagneBike [5-8],
(c) MagneBike PCAE [9]**

5.2.3.1 Related prototypes and state of the art in 2005

At the beginning of the project, a detailed survey on the state of the art had been performed, mainly stressing on classical in-pipe robots that spread within the pipe (also see 3.1.1.3 and Fig. 5-9-a). As the difference between the smallest diameter in the access pipe and the biggest one within the steam chest is mostly bigger than factor 3, and also the diameter changes are often very abrupt, it soon became evident that commercially available robots of this type do not achieve the requested mobility. More sophisticated concepts for in-pipe-robots that actively spread using several DOF (e.g. the Moritz [55]) had to be rejected as well due to their high mechanical complexity which normally leads to a size that does not fit through the Ø250mm access pipes any more.

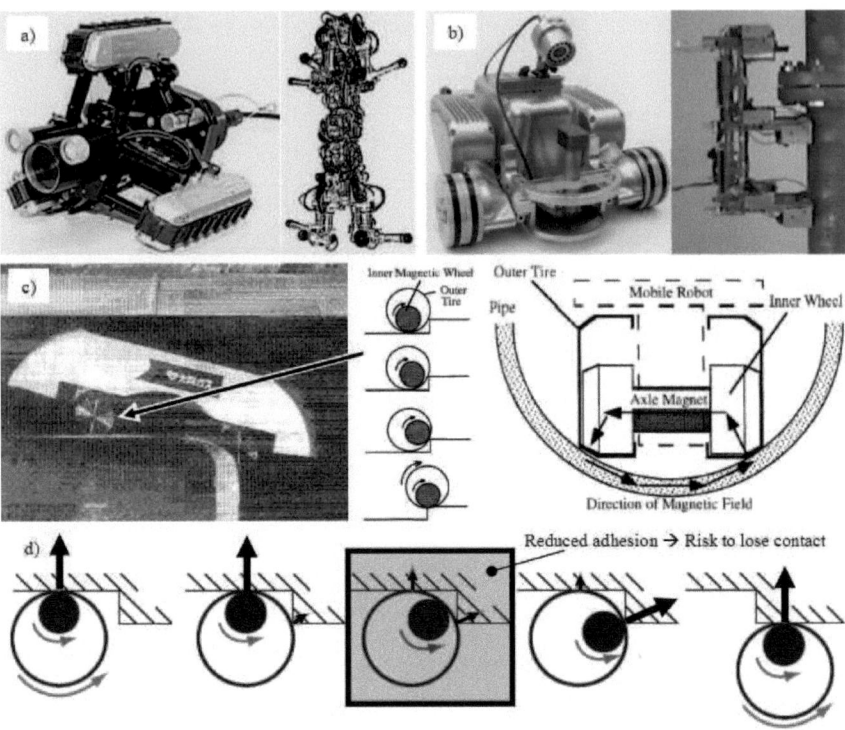

Fig. 5-9: In-pipe-inspection robots and magnetic wheeled robots that already existed before the project start (year 2005)
(a) In-pipe robots that mechanically spread (MicroTrac Vertical [56], MORITZ [55]), (b) Magnetic wheeled robots (Tripod [31], PIR [48]),
(c) Sewage-pipe inspection robot by Osaka Gas [40], using the "dual magnetic wheel" for passing inner transitions and steps on the ground [41],
(d) Limitation of the dual magnetic wheel on steps on the ceiling

For this reason, it was decided to take advantage of the ferromagnetic material and to focus on a magnetic wheeled robot at a size that is relatively small compared to other in-pipe inspection robots with similar mobility (Fig. 5-9-a, b). As already described in the case study about the gas-tank-project, magnetic wheeled robots with enough mobility on the specified obstacles and small enough for the access pipes of only Ø250mm did not exist at this time yet: Simple robots such as the Tripod by Jireh industries [31] cannot pass the complex combinations of inner and outer transitions, while more sophisticated designs (PIR [48] and our first concept for gas tanks [1]) are too big and complex.

5. Case studies   121

The only magnetic wheeled robot at this time that was optimized for the inner sections of pipes with abrupt diameter changes was the sewage pipe inspection robot by Osaka Gas ([40] and Fig. 5-9-c). It uses the "dual magnetic wheel"-mechanism [41] for passing inner transitions and small steps on the ground. As already explained in the chapter about the mechanisms for inner transitions (4.2.1.4), the "dual magnetic wheel" is only suitable for steps on the ground, as the adhesion force gets decreased significantly in the middle of the transition. This reduced adhesion force would cause the wheel to lose contact on a vertical wall or on the ceiling (see Fig. 5-9-d) – resulting in the robot falling down.

5.2.3.2 Calculation model and first test prototype

For better understanding the behavior of magnetic wheeled robots in complex combinations of inner and outer transitions, a 2D mechanical calculation model was established. With this model and realistic values for the magnetic adhesion force, the robot mass and the geometrical values, the minimum required friction coefficient between wheel and surface (μ_{min}) can be calculated. For estimating the risk of slipping, this value can then be compared against the measured value of real wheels – which normally ranges from approximately μ_{real}=0.2-0.3 for blank steel wheels up to μ_{real}=0.8 for wheels with good rubber cover.

The basic method how to calculate the value for the required friction coefficient can be seen in Fig. 5-10-a – at the example of a basic 90° inner transition that is passed with a 2-wheeled vehicle without additional mechanisms. As already outlined in the chapter about this specific type of obstacle (4.1.1.1), the worst case with the highest need for traction occurs when one wheel gets detached from the old surface (Fig. 4-1-b) – which needs to be calculated for all wheels of the vehicle. For getting the values for all unknown forces (F_{T1}, F_{R1}, F_{T2}, F_{R2}), a matrix-equation can be established out of the 3 equations for force- and moment-equilibrium and an additional equation for the torque-distribution; and solved with the "\"-operator in MATLAB or a similar program. The required friction coefficients and torques in all wheels can then be calculated out of these force-values. Concerning the equation for the torque-distribution between the two wheels, "a_2/a_1" = "T_1/T_2" can either be set to "1" (which corresponds to equal torque in both wheels) or an iterative process can be used where it is set to the ratio between the friction coefficients (μ_1/μ_2) calculated in a previous iteration. After a few iterations, the μ_1 and μ_2 usually converge to the same value.

5.2. Steam chests

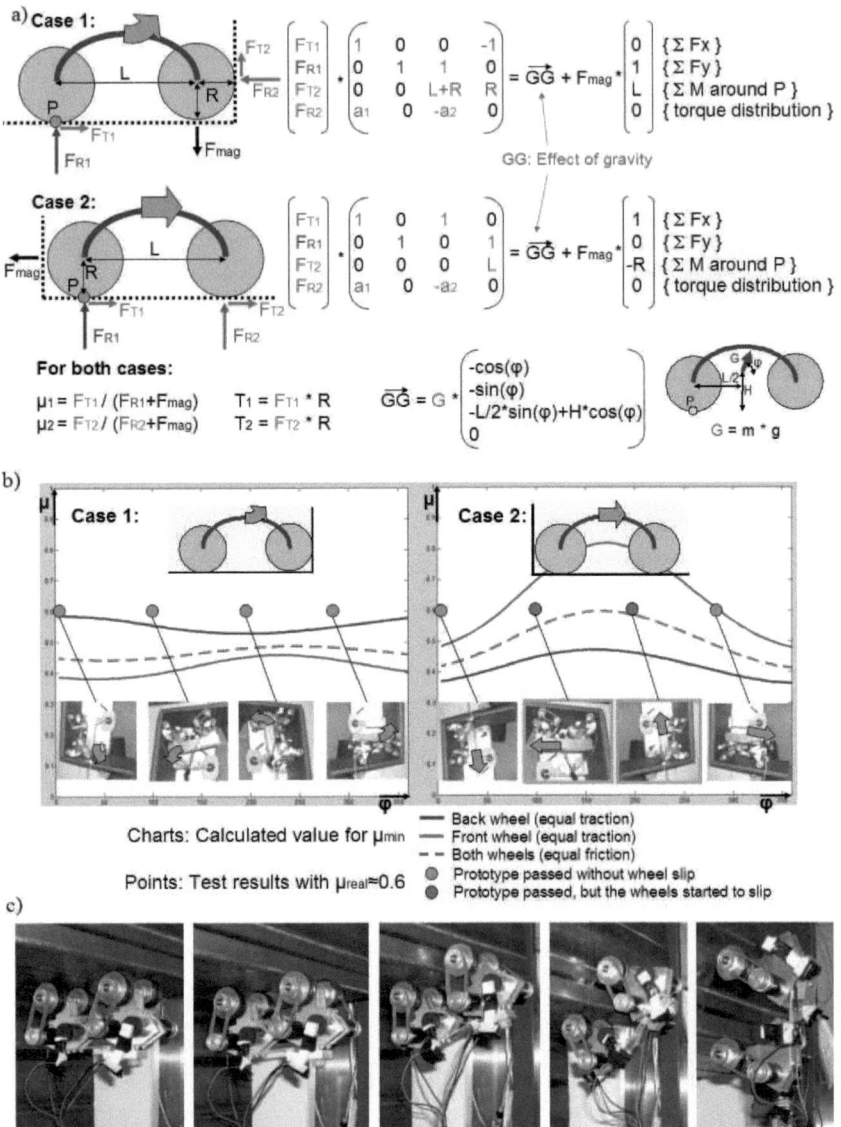

Fig. 5-10: Quasi-static calculation model for passing inner 90° transitions and its verification in a simple test prototype [3]
(a) Derivation of the matrix-equation, (b) Calculation vs. test results,
(c) Passing an outer transition with the test prototype

5. Case studies

The calculation results for a simple vehicle with only 2 wheel-pairs and all-wheel-traction, but no additional mechanisms for obstacles implemented, is represented in Fig. 5-10-b – using the values for mass, magnetic force and geometry from a test prototype that was built for the verification of this calculation model. As it can be observed in these graphs and as it was also proven in real experiments, such simple robots are already able to pass inner transitions in all possible inclinations without using additional mechanisms. However, in some cases the limit of slipping is already very close – which means that on wet surfaces or by using more robust wheels of lower friction coefficient it cannot be guaranteed that the robot is still able to pass. Also on steps and other combined obstacles that can be found in steam chests such simple vehicle structures can normally not pass any more. More details about the calculation model and the detailed design of the prototype can be found in our paper at CLAWAR08 [3]. More complex calculations for steps and other combined obstacles are described in the student report of Marco Morales [4].

Similar to the MagneBike, also this first prototype is also able to pass outer transitions (edges) with no significant force decrease on the edges – as it can be seen in Fig. 5-10-c. However, sharp edges cannot be passed either.

5.2.3.3 MagneBike

For achieving the required mobility also on multiple steps, holes and transitions on curvatures, the MagneBike [5-8] was developed.

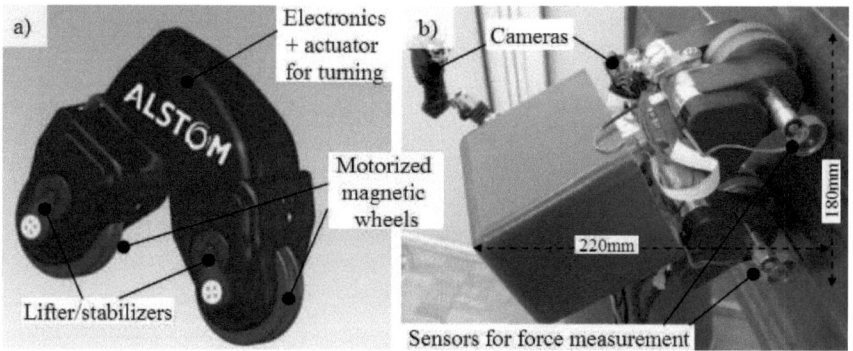

Fig. 5-11: The MagneBike [5-8]
(a) First design-study for showing the basic concept with the main components (b) Prototype implementation with cameras, sensors and the first generation of the electronic boards

As it can be seen in Fig. 5-11, this robot is composed out of two wheel units with one motorized magnetic wheel and two non-magnetic extra wheels each; and a main body with an actuator for turning the front wheel unit. The extra wheels are fixed on a shaft that can rotate coaxially to the rotation axis of the magnetic wheel – which allows for two functions at the same time:

- Local force reduction by using them as rotary lifters (see Fig. 5-12-a), which allows for mobility on inner transitions, steps and other combined obstacles without negative side-effects on the edges of outer transitions.

- Active stabilization on traversing paths on vertical walls (see Fig. 5-12-b). In contrast to just passively stabilize by using two main wheels per unit, this approach allows for building very compact and to well adapt to surfaces that have a very small curvature radius in respect to the robot size.

In addition to the good adaptation in curved surfaces, the concept of using only two main wheels requires only 5 active DOF in the entire vehicle for steering and turning on spot (see Fig. 5-12-c) – which is a relatively low number compared to previous prototypes with similar mobility.

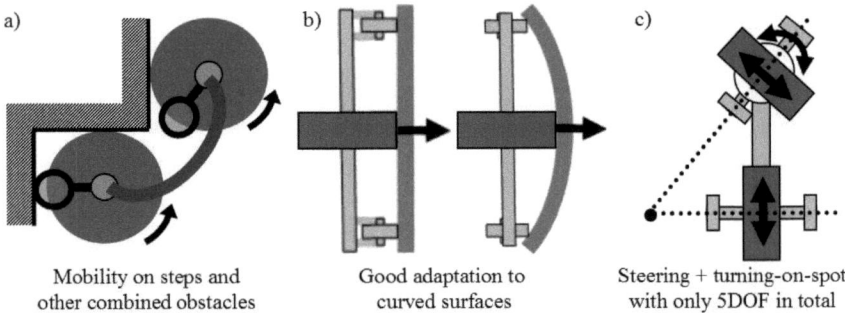

Mobility on steps and other combined obstacles

Good adaptation to curved surfaces

Steering + turning-on-spot with only 5DOF in total

Fig. 5-12: The three basic conceptual ideas in the MagneBike concept and their core advantages
(a) Rotary lifter (patented technology [8]), (b) Active stabilization by using the same arms as for the rotary lifter, (c) Vehicle structure in bike-configuration which allows for steering with few active DOF

For transmitting torque to the two coaxial shafts of wheel and lifter in a very compact way, the gear configuration that is represented in Fig. 5-13 resulted the most promising: Two DC-motors with planetary gearboxes are mounted above the wheel – parallel to its rotation axis and with their output shafts pointing in opposite directions. On each side, the torque is transmitted to the main axis with a double-stage spur-gear transmission. The second stage of this transmission is on the inner side of the structure, which allows for fixing the last gear of the wheel-motorization directly on the rim of the wheel. For symmetry-reason, this design is also realized for the transmission to the lifter shaft. In addition to just transmitting the torque to the final destination, the spur-gear-transmission is also used for implementing an additional reduction factor of approximately 1:4 – which allows for using more compact actuators with less torque.

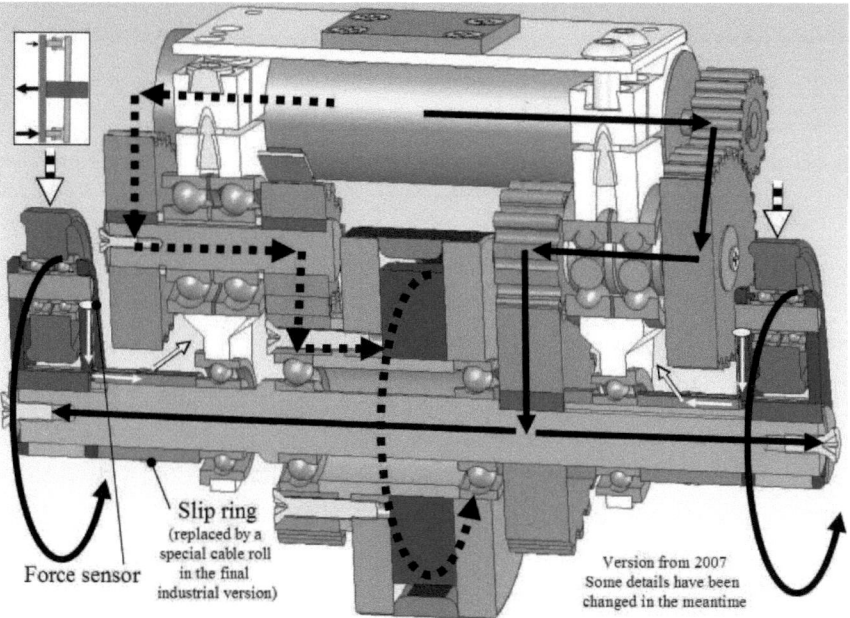

Fig. 5-13: Cut-view of a wheel unit in the MagneBike, showing the torque transmission for the wheel (black dashed line) and the lifter/stabilizer (black normal line); and the signal transmission for the two force sensors on each extra wheel (white lines) that are necessary for stabilizing the robot on a traversing path along a vertical wall (small image)

Note that the necessary torque for turning the lifter-arms is quite high in relation to the small vehicle size – 7.7Nm. The need for such a high torque basically has two reasons: At first, the magnetic wheels need to be very strong for assuring a safe operation also on sharp edges in environments with low-ferrite steel alloys – where the force can get decreased down to approximately 25% of its original value (~50% reduction caused by the low-ferrite steel, multiplied by 50% caused by the saturation-effects on the edge). For this reason, the wheels are designed with a maximum adhesion force of 250N each – which is more than seven times the gravity force coming from the robot mass (3.5kg x 9.81 N/kg = 34.3N). As the rotary lifters do not only have to be applied on flat ground, but also on curved surfaces, in the worst case (smallest pipe of only Ø250mm) the length of the torsion arm for the lift movement reaches up to almost the full length of the lifters (26.8mm) – which then results in the maximum required torque of 7.7Nm (= 26.8mm x 250N). For transmitting this high torque at a reasonable safety against breaking the teeth of the gears, it was necessary to use the full available width and the highest available steel quality for the gears.

Besides the need for high torque at little available space, the main challenge in the detailed design of this robot was the force-measurement of the reaction forces on the extra wheels. For moving on traversing paths in vertical walls (see Fig. 5-13 small image), the information about these forces is necessary for controlling the lifter/stabilizers. Without this control, the stabilizers would either not stabilize well or detach the main wheel unwanted. In the context of force-measurement and stabilization, the following sub-problems had to be solved:

- Implementation of a control algorithm for the stabilization that is fast, robust and executable on the onboard-processor of the robot.

- Tools for a simple calibration of the sensors.

- Measurement both on plain surfaces and in small pipes, which results in a change of the attack-angle for the force of more than 90° (cannot be realized with a 1-dimensional sensor any more).

- Robust signal-transmission to the robot through a rotating shaft.

After several iterations, these challenges could finally get solved in a satisfying way – using elastic elements with customized strain gages for the force measurement and sophisticated algorithms for control and calibration. The originally planned slip-rings for the signal transmission were replaced by a special cable-roll in the final industrial version; to assure a higher robustness.

More information about the stabilizer function, solutions for other challenges in the design and a report on detailed tests with the robot and its sub-components can be found in our paper at JFR [7] and in the PHD-thesis of Fabien Tâche [105]. A photo of the prototype is represented in Fig. 5-11-b, and the most important mechanical properties are listed in Table 1.

Parameter	Value
Size[a], $L \times W \times H$	$180 \times 130 \times 220$ mm^3
Height of center of mass z_{CM}	65 mm
Wheel distance L_w	120 mm
Wheel diameter $2r$	60 mm
Mass m	3.5 kg
Mass repartition	Wheels: 23%, actuators: 18%, gears: 17% structure, housing, electronics: 42%
Maximum magnetic wheel force F_{mag}	250 N (NdFeB magnets)
Wheel torque $T_{w\_cont/int}$	2.1 Nm (continuous), 6.7 Nm (intermittent)
Lifter torque Tl_int	7.7 Nm (intermittent)
Steering torque $T_{steer\_cont/int}$	2.33 Nm (continuous), 4.1 Nm (intermittent)
Operating voltage	24 V (actuators), 5 V (electronics)
Power (at max. speed)	4.6 W (horizontal), 6.7 W (vertical)
Communication	RS232 at 115,200 baud
Maximum speed v_{robot_max}	2.7 m/min
Steering rotation speed w_s	33 deg/s
Control mode	Remote control with onboard motor controllers

[a]Will be smaller than 200 mm with fully integrated electronics.

Table 1: Main mechanical properties of the MagneBike robot [7, Table 1]

Thanks to the important economic value of the steam-chest application and the great support from ALSTOM, the industrialization has already started with a small series of 5 units. The final industrial design of this prototype was mainly done by Tresch&Kielinger engineering for the mechanical parts and IfTest for the electronic components – with support from our team and ALSTOM Inspection Robotics. Thanks to the expertise of all involved partners during the last phase, these prototypes are now robust enough for the end-user.

However, during the tests also some limitations were noticed:

- Even with the newest control algorithm for the active stabilization, the maximum allowed speed on paths where this stabilization is necessary gets slowed down significantly (less then half of the normal speed, with the control programs used in 2009) – which means an increase of the total inspection time.

- With the active stabilization, tilting on double-curved surfaces cannot be avoided (reduced adhesion force on such surfaces, see Fig. 5-14-a).

- An even stronger force reduction occurs on sharp edges. Tests with the robot carrying different payloads showed, that on a sharp edge environment the payload was only 3kg, while in normal operation up to 12kg could be carried. Even if the remaining 3kg on these edges still seem very high, it has to be mentioned that these tests were performed in a clean environment made out of standard steel and without curvature – while in real steam chest the expected forces could be lower and already reach a critical value on sharp edges.

- More difficult outer transitions or combined obstacles (e.g. thin ridges in gas tanks; see previous case study) cannot be passed with the MagneBike. Even if not necessary for moving in the originally specified steam-chests, extending the application scope of future robots to these more difficult environments was regarded as a promising research goal.

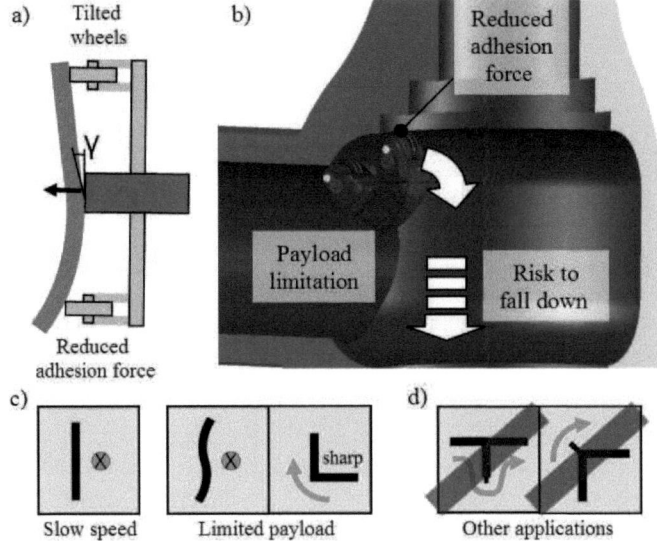

Fig. 5-14: Main limitations of the MagneBike
(a) Problems on double-curved surfaces, (b) Reduced adhesion on sharp edges and its consequences, (c) Environments in steam chests where the performance is slightly limited, (d) Obstacles in more difficult environments (e.g. gas tanks) where the MagneBike cannot move

5.2.3.4 MagneBike "PassiveCorner Active Edge" (MagneBike PCAE)

For addressing these above-mentioned limitations, the MagneBike PCAE was developed. As it can be seen in Fig. 5-15, this new concept reuses several components, ideas and technologies from previous designs: The steering-unit and all electronics from the MagneBike, the WpW-technology (4.2.1.3, [10], originally developed for robots in generator housings) and a vehicle structure with active joints, similar to the first concept developed for the gas tank project [1].

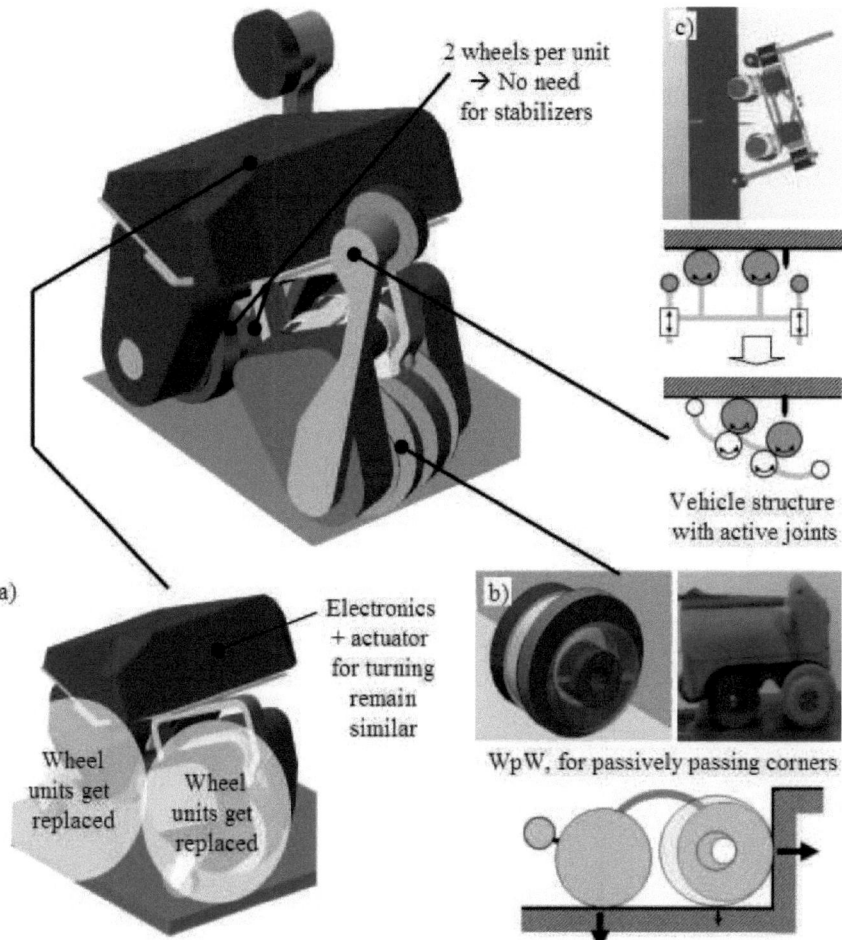

Fig. 5-15: CAD-model of the MagneBike PCAE [9], showing how components, ideas and technologies from previous prototypes are reused (a) Original MagneBike [7], which uses the same upper part with electronics and the actuator for turning, (b) Wheel-parallel-to-the-wheel-technology for passing corners, originally developed in the context of the generator housings [10, 14], (c) Vehicle structure with active joints for passing sharp edges, ridges or other difficult combined obstacles – similar to the first concept developed for gas tanks [1]

With this combination of innovative ideas, concepts and technologies, the following advantages can be achieved:

- By using the Wheel-parallel-to-the-wheel (WpW) technology for passively rolling through corners/inner transitions (Fig. 5-15-b) there is no need for using an additional DOF for powering the rotary lifter on this type of obstacle.

- For avoiding the need for active stabilization (which in the original MagneBike is realized with the same arms as for the lift-movement) the wheel units in the MagneBike PCAE use a pair of 2 main wheels instead of only one and a differential gearbox in between to keep the torque for turning at a low value. With this configuration, the robot can move faster on traversing paths in vertical walls, as the time-consuming control of the stabilizers is not necessary any more. In addition, one active DOF per wheel unit is saved. The drawback of this approach is a slightly decreased adhesion force on curved surfaces (tilting effect, see Fig. 4-7-a). However, this force decrease only gets critical on curved edges – where it does not cause any negative effect any more if the active-edge-extension-arms are used (see next paragraph).

- In the wheel units of the MagneBike PCAE, the "saved" active DOF which in the original MagneBike powers the lifter/stabilizer-arm is used for moving the active-edge-extension-arm (Fig. 5-15-c). With this arm, the robot is able to assure the contact of a wheel that is moving over a sharp edge, a thin ridge or any other zone that does not provide enough adhesion – and thus increases the application scope to these environments (e.g. gas storage tanks, see previous case study). In addition, the payload capability on "normal" sharp edges gets increased as well, as the force reduction there (= main payload limitation for the original MagneBike) does not get critical any more.

- By only replacing the wheel units and reusing the upper part of the original MagneBike (actuator for steering, electronics), the total production and development costs are decreased significantly (estimation: <20% of total cost). Because of the high compatibility to the old design, the MagneBike PCAE is not defined as a "new prototype" but as an "extension" or "upgrade" for the existing system.

The first iteration for the detailed design of the wheel units has been realized in a student project by Lorenzo Bagutti [9], where the first generation of wheel units was realized by using 3D-printed plastic parts for structure and gears; and magnetic wheels and motors from old prototypes. Starting from this design, the second version was derived. A cut-view of the second design for the wheel unit is represented in Fig. 5-16.

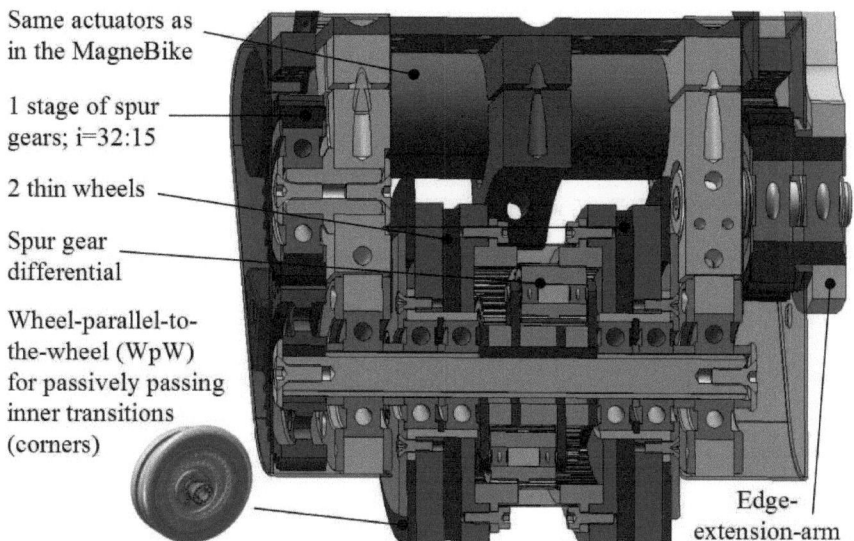

Fig. 5-16: Cut-view of a wheel unit in the MagneBike PCAE (2nd design), showing the wheels with WpW-technology, the torque-transmission, the spur gear differential and the edge-extension-arm

For powering the wheel shaft and turning the edge-extension-arm, the torque from the motor is transmitted via a one-step spur gear transmission with an intermediate wheel (reduction factor i=32:15). Compared to the old design, this solution only provides half the torque (3.8Nm instead of 7.7Nm) but allows for more space close to the wheel – which is necessary for placing two magnetic wheels with discs for the WpW-technology and a differential gearbox in between. Tests at smaller size showed that corner-passing with a wheel that uses this new technology needs approximately 1/3 the torque that is necessary with the current rotary lifter (torque arm: 26.8mm, similar to wheel radius: 30mm). For this reason, the lower gear reduction of only 32:15 instead of 4:1 is enough. For powering the edge-extension-arm, a torque of 3.8Nm was calculated to be sufficient as well. While in the first prototype during the student project only plastic gears were available, the final version uses high quality steel gears similar to the ones in the original MagneBike. For saving mass, alternatives solutions for the torque transmission (gear-belt for the wheel shaft, wire-transmission for the edge-extension) have been tested as well, but did not result in robust and reliable solutions.

As the original steering unit should be kept, the torque for turning the wheel pair needs to stay at a low value. For this reason and for decreasing the slippage, a differential gearbox is included. As standard differential gears with bevel wheels could not be found at this very small size, a customized spur gear differential was developed and could transmit the required torque already in the first version that was only manufactured out of cheap 3D-printed plastic gears.

For facilitating future work on optimizing the magnetic wheels and the WpW, the wheel unit is designed in a way that allows for disassembling it with little effort. In order to measure the force on the edge-extension-arm (necessary for a fully-autonomous control), the same force sensors are used as in the lifter arms in the old design. When the robot is applied in an environment with no sharp edges, the edge-extension-arms can be removed without effort.

5.2.4 Innovations, outlook and lessons learned

5.2.4.1 Innovations

Both the original MagneBike and its upgraded version, the MagneBike PCAE consist of two innovations each – one in the obstacle-passing-mechanisms and one in the vehicle structure. Given the expected benefits in the highly valuable business of steam chest inspection, 2 out of these 4 innovations have been patented at ALSTOM and one at least included as a group of sub-claims. The active rotary lifter was already patented in 2007 [8]. The application process for the wheel-parallel-to-the-wheel [10] was still ongoing at the submission time of this thesis (May 2010). In that patent application, the vehicle structure of the MagneBike PCAE [11] is included in the sub-claims. Also the original MagneBike structure [6, 7] with active stabilization was new and looked advantageous at the time it was developed – even if it was finally not patented.

5.2.4.2 Industrialization and outlook

Five units of the original MagneBike are already produced and will be used both for inspection tasks in the field, but also as a research platform for future projects on localization, navigation and coverage in pipe-like quasi-3D-environments – which aim for fully autonomous inspections in the near future. Starting with the works of Fabien Tâche in the field of localization and mapping by using a compact laser range finder; this research will mainly be continued in collaboration with ALSTOM and Massachusetts Institute of Technology under the responsibility of Andreas Breitenmoser who will do his PHD thesis on multi-robot-systems and coverage with several units of the MagneBike.

Concerning the PCAE-wheel units, the upgrade of one MagneBike-prototype was still ongoing at the submission time of this thesis – with the option to upgrade the rest as well, if the tests are successful. By also keeping some of the prototypes unchanged, there will be a modular fleet of 3 almost similar robots with high compatibility within the fleet but specific advantages in all of them, as it is represented in Fig. 5-17.

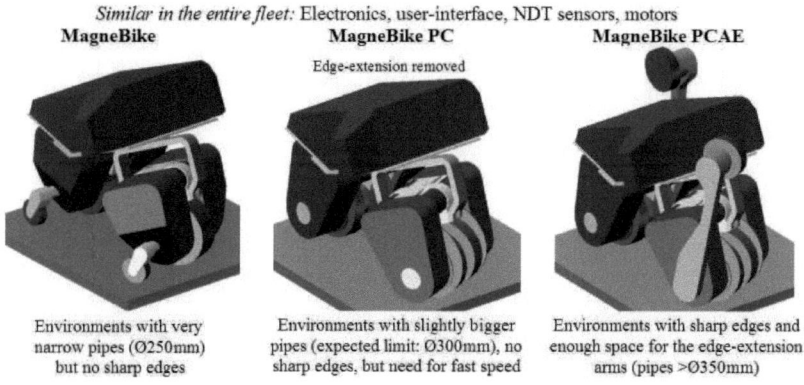

Fig. 5-17: Modular fleet within the MagneBike-robots, pointing to the specific advantages and applications of all three types

5.2.4.3 Conclusion and lessons learned

What can be learned from this project is the finding that even a sub-system that looks simple and highly innovative at first glance can become very complex and significantly increase the cost of a project. This was the case of the active stabilization in the original MagneBike-structure, which finally led to several months of unplanned work, still forms a minor limitation of the system and is replaced in the next generation. Even if sensing and control was integrated into the design process right from the beginning and a highly competent steering-committee re-asked all decisions every few months, neither the unexpected difficulties with the force-measurement nor the existence of more promising concepts could be foreseen – at least not in 2006. With the now available knowledge about more than 10 different mechanisms for corner-passing (chapter 4.2.1, especially Fig. 4-23) and the experience about force-measurement and control (PHD Fabien Tâche

[105]), future research teams will have a much better knowledge-base for planning the effort of similar projects.

Another conclusion that can be drawn from this project is the observation that the success of an industry-related project in applied research mainly depends on the willingness-to-pay of the industrial partner. For this reasons, the steam-chest project finally got significantly more successful than the one on gas-tanks – even if the environment challenges in this applications were less difficult and the technical quality of innovations in the MagneBike and the first concepts on gas tanks are comparable. The key factors for achieving such a high-willingness-to pay were the following:

- Collaboration with an industrial partner that can assure both a high financial support and is open for new and innovative technologies. As a general tendency, in big international companies these constraints usually are fulfilled better than in small- or medium-sized ones, as the available budget for R&D is usually much higher and the responsible project managers normally hold a PHD in engineering and thus like to support projects similar to their own thesis in the past.

- Professional product presentation with a strong focus on industrial design. For first presenting the concept of the MagneBike, lots of effort was spent on a computer animation and a design-prototype with a nice housing in the corporate colors of ALSTOM. Both the animation and the photos of this design prototype caused very positive reactions at the most important decision-makers within the management – and assured the funding of both this and the follow-up projects already before any prototype was built.

What can also be realized in this context is the observation that the value of an innovation can get significantly increased as soon as its application scope gets extended to a high-value business-case: The wheel-parallel-to-wheel technology did not gain much attention at the time of its invention, because it had originally been developed for the generator-housing-scenario – an application with relatively low economic significance. Only by extending its application scope to the highly valuable business-case of steam-chest inspection, the economic value could be raised high enough for filing a patent.

5.3 Generator back housing

5.3.1 Motivation and business relevance

As already explained in the first chapter of this thesis, generators are not only one of the most expensive components of all, but also highly stressed due to vibrations, high currents and elevated temperatures. For this reason, their inspection – especially the one of the stator winding (next section) – is one of the most important business cases in power plant service. Also the back housing has to be accessed from time to time – not only for searching cracks or other damages, but mainly for installing vibration sensors. These sensors are necessary for measuring unwanted irregularities that decrease the lifetime of the generator.

Also many other components in power plants, ships or refineries use structures with a geometry that is highly similar to the one encountered in generator housings. For this reason, a universal robot at very small size (around factor 2-3 smaller than the MagneBike = max. 120mm height) that can deal with the most difficult challenges in this type of environment was seen as a very useful tool for such alternative applications.

5.3.2 Environment specifications

As it can be seen in Fig. 5-18, the basic geometry-constraints in generator housings are relatively similar to the ones in steam chests – ferromagnetic environments with inner and outer 90° transitions and little space.

However, there are some significant differences in terms of size-restrictions and complexity of obstacles:

- Size + curvatures: No need to adapt to very small curvatures (Minimum curvature radius: 1500mm instead of 125mm), but harder size restrictions (100mm x 120mm instead of a pipe with a minimum diameter of Ø250mm)

- Obstacles: No need to pass tripple steps or holes, but 90°-corners and a new worst-case-obstacle: Ridges with 20mm width.

5.3. Generator back housing

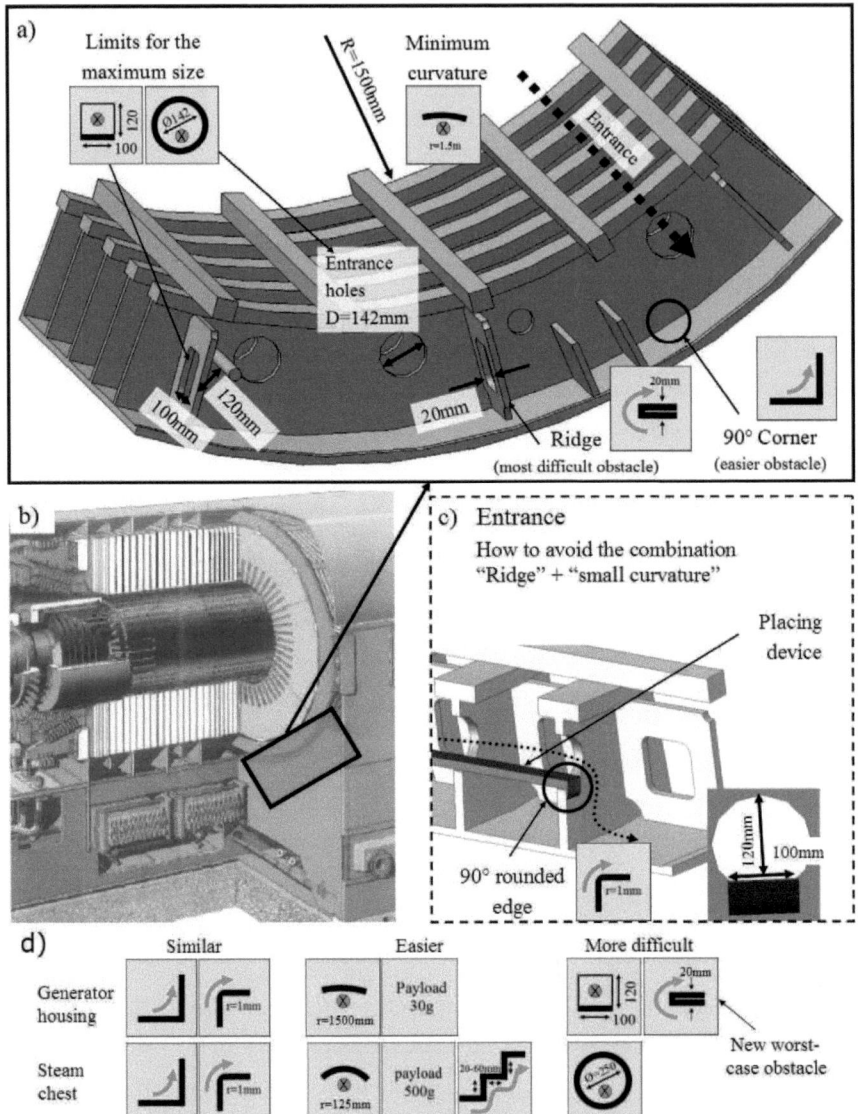

Fig. 5-18: The generator-housing environment [12]
(a) CAD model showing the most difficult obstacles and size restrictions,
(b) Position within the generator, (c) Method of facilitating the entrance,
(d) Comparison to the specifications in steam chests (previous case study)

5.3.2.1 Limitation on the wheel axis distance

As already pointed out in the first case study about the gas-tank-project, mobile robots that are able to pass ridges in all possible inclinations of gravity normally result in relatively complex vehicle structures with active joints (see Fig. 4-25). Among these vehicle structures, even the simplest one still needs 4 DOF without the mechanism for steering included and results in a relatively large prototype. In the case of the first robot for gas-tanks [1], its size is already 200mm in the smallest direction and its mobility does not allow for passing 90° inner transitions (corners) that are not part of the ridges. Downsizing such a mechanism with approximately factor 2 and improving its mobility at the same time seemed out of scope in the context of this project.

Instead of following such an approach, we realized that the difficulty of an obstacle strongly depends on its size relatively to the robot (see chapter 4.1.6 and (Fig. 4-11). A ridge remains a "difficult obstacle" if the robot is relatively big compared to its size and requires a rather complex mechanism to pass it. In contrast, for a relatively small robot, it is just a combination of 90°-corners and 90°-edges – which can be passed with a relatively simple vehicle structure. The limit condition for completely separating a "ridge" into two "edges" and two "corners" is reached, when the wheel axes distance (L) is smaller or equal than the width of the ridge (b=20mm in this application).

5.3.2.2 Additional specifications

Apart from the limits concerning size and mobility on certain obstacles, the payload had to be defined as well. For carrying at least a camera plus the vibration sensor, this requires 30g and a space of 20x20x30mm^3.

To guarantee a long lifetime and a reasonable resistance against abrasion of the wheel rubber, a rather hard and resistant rubber had to be used. As such hard rubbers do not achieve the same friction as soft ones and also the surface is sometimes dirty, we defined a minimum required friction coefficient of $\mu=0.5$.

As non-mandatory options we also defined that the robot should be able to pass the 20mm-ridges on the entrance holes with small-curvature (D=142mm, see Fig. 5-18) and to move on curved surfaces that can be found in steam chests (D=250mm) or other environments in power plants. This enlarges the application scope of the robot. As a cable to the robot was anyway planned for safety reasons (emergency-removal in case of a total failure) and for transmitting the signals from the vibration sensor, "power autonomy" and "wireless communication" were only put as non-mandatory options for potential other applications.

With all these inputs, the following specification list could be derived. Note that passing inner and outer 90°-transitions at all possible inclinations, with a minimum required friction coefficient below 0.5, had not been realized with any vehicle of the required size (L<20mm) yet at the time of the project start (2006).

Mandatory (M) or optional (O)	Description	Value
	Mobility on specific geometries / obstacles	
M	All inclinations of gravity, // and ⊥ to driving direction	0° - 360° *
M	Inner transitions (corners)	90° +/-5°*
M	Outer transitions (edges)	90° +/-5°*
M	Minimum friction coefficient wheel - surface (determined by the	μ = 0.5 *
M	Concave curvatures, // and ⊥ to drive direction	R = 1500mm
O	Concave curvature in combination with edges	D = 142 mm
O	Concave and convex curvatures, // and ⊥ to driving direction	D = 250mm
	Size limitations	
M	Max. width	B = 100mm
M	Max. height	H = 120mm
M	Max. wheel axes distance – to separate a ridge into 2 edges	**L = 20mm ***
M	Enough ground clearance for 90° edges and 90° corners	
O	Enough ground clearance for curved surfaces	R = 71mm
	Payload and power supply	
M	Space for the payload	20x20x30mm³
M	Mass of the payload	30g
O	Power autonomy + wireless communication to the user	

*:) Specifications that are the most difficult to fulfill with previous vehicle structures

Table 2: Specification list for the mechanical design of the robot for generator housings [12]

5.3.3 Prototypes

In contrast to the steam-chest project (previous case-study), which was done in the context of a smaller project with only our team at ETH involved, the project on generator housings was done in the framework of the nation-wide project KTI 8435.1 EPRP-IW – with also researchers from EPFL developing prototypes. Additionally, teams from other research institutes developed similar prototypes at the same time as well. Even if none of these external prototypes fulfills the required specifications in generator housings, there are some common aspects and interesting partial solutions. For this reason, the most remarkable ones are described and compared against our prototypes in the last subchapter.

Fig. 5-19: Prototypes developed for the back housing of generators (a) First prototype with passive extra wheels in the structure for easier passing inner transitions [12], (b) Bigger version of this robot realized in the MSc-thesis of M. Oeschger [13], (c) Rotating magnetic cam disc [15], (d) Modular prototype for testing alternative mechanisms for passing inner transitions, (e) Micro-Tripod WpW [14], (f) Remarkable prototypes from our project partners at EPFL and other research teams, with comparable size and mobility (Tripillar [39], First prototype with rotating adhesion zone [44], CyMag [45], Gecko [46], WaalBot [84], Gel type sticky mobile inspector [85])

5.3.3.1 First prototype with passive extra wheels in the structure

As already described in chapter 4.2.1.2, vehicle structures with additional non-motorized extra wheels in the structure allow for passing concave corners at a significantly reduced necessary friction coefficient between wheel and surface and thus can also be used in combination with hard rubber and/or wet surfaces. At the time of the first prototype design for this application (end of 2007), this mechanism was the only one that allowed for designing the vehicle small enough so that the ridges are completely separated into two outer 90°-transitions – with the wheel axis distance L being 20mm. An overview over the mechanical design and the core components of this prototype is represented in Fig. 5-20.

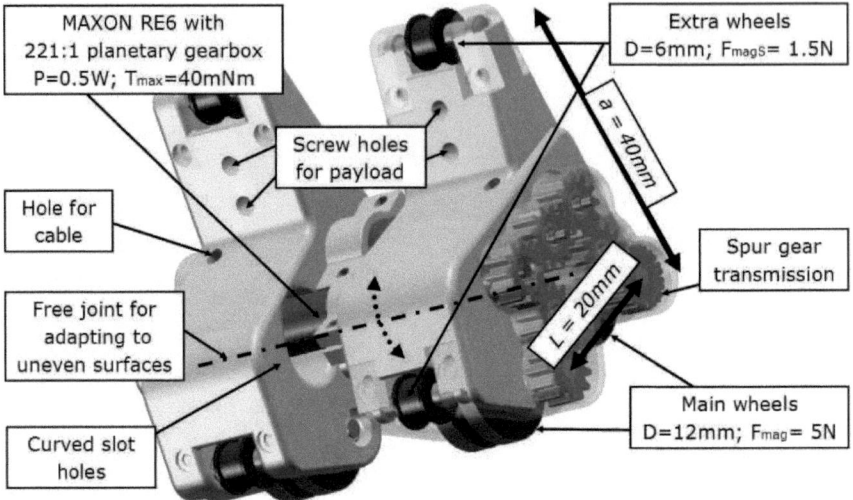

Fig. 5-20: Mechanical design of the first robot with additional non-motorized extra wheels in the structure [12]

For choosing reasonable values for the core parameters, mechanical calculations based on quasi-static 2D-models have been performed – similar as already done for the preliminary test prototype for the steam-chest-scenario with only two motorized pairs of wheels (see Fig. 5-10-a and [3]). In contrast to the calculations for that other vehicle structure on only 2 wheel pairs, the model for the here analyzed structure (4 wheel pairs) consists of 4 instead of only 2 cases (each wheel needs to be detached).

Based on these calculations and the width of the ridges (b=20mm) the final values for the robot geometry and the magnetic force in the wheels were defined. An example for the derivation of one of the equations is represented in Fig. 5-21. The complete calculation model and the derivation of the equations for the central design parameters can be found in our paper at IROS09 [12].

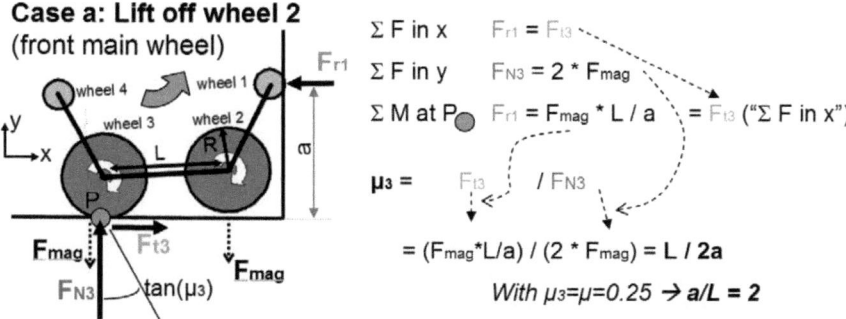

Fig. 5-21: Example for the derivation of the equations that are necessary for defining the core parameters. The entire calculation model can be found in our paper at IROS09 [12]

This prototype was successfully tested both in an artificial lab environment and in a steam chest (Fig. 5-22), as the obstacles in this environment are highly similar as in the generator housings.

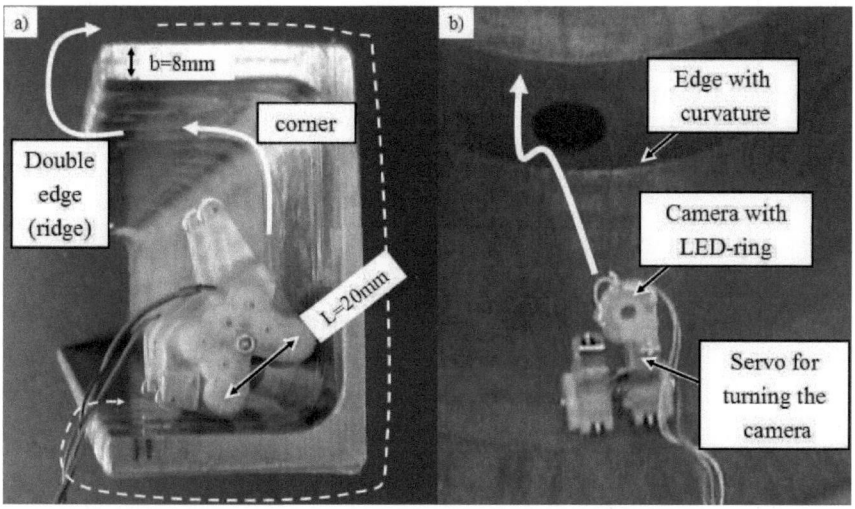

Fig. 5-22: First prototype for the generator housing, with passive extra wheels in the chassis to better move through inner 90°-transitions [12]

During the tests, we realized that the robot had some problems with small rust particles and that the torque provided by the motor was only high enough to pass corners on painted surfaces (as specified) but not on unpainted surfaces (magnetic force too strong). Small bumps in the steam chest caused severe problems to this first prototype. On the other hand, the ridge-passing ability was much better than expected: The limit was not b=20mm (=L) but only b=8mm (=L-2R).

Driven by these test results and the fact that the payload capability of the first prototype was not very high (approximately 10g), the next version was realized at bigger size. For avoiding that the motor of one wheel unit penetrates into the other one (as it is the case in the first version), the planetary gearboxes in the first prototype are replaced by a worm-gear-transmission integrated into the robot chassis (Fig. 5-23). More technical details about the design can be found in the MSc-thesis of M. Oeschger [13].

**Fig. 5-23: Bigger version realized in the MSc-thesis of M. Oeschger [13]
(a) Symmetric design with 4 extra wheels, (b) Self-made worm-gear-transmission, (c) Photo of a disassembled wheel unit**

5.3.3.2 Low-cost prototypes for testing alternative corner-passing mechanisms

Although the first two prototypes already fulfilled all mandatory requirements regarding mobility and size restrictions, the search for even simpler solutions remained interesting from the scientific point of view. Inspired and motivated by results from our partners at EPFL and other research teams at the same time, new concepts for such mechanisms have also been developed in our team at ETH; and successfully proven in low-cost test prototypes.

The simplest of these mechanisms is the rotating magnetic cam disc [15]. Mainly inspired by other climbing robots rolling on whegs (e.g. the WaalBot [84]), this concept takes advantage of the observation that wheels that are not round can pass inner transitions very easily – thanks to the inhomogeneous distribution of the adhesion force. Preliminary tests with 2 screw nuts, a ring magnet in between and a torque transmission through a flexible shaft have been performed – leading to the following basic measurement results:

- By turning the flexible shaft, the magnetized screw nut can passively roll through inner transitions without any problems. This ability can be achieved without any rubber cover on the screw or the steel surface ($\mu \approx 0.3$), whereas a normal wheel would get stuck.

- The required torque for turning a magnetized screw nut with 7mm wrench size is approximately 20mNm. This value is higher than turning a normal wheel of similar size, but still low enough for using the same gear motors as in the first prototype (Maxon RE6 with 221:1 gearbox, 60mNm).

- The ratio between the adhesion force on the flat and on the sharp side of the nut is approximately factor 3, with the lower value on the sharp side still comparable to normal magnetic wheels of similar size (approximately 10N).

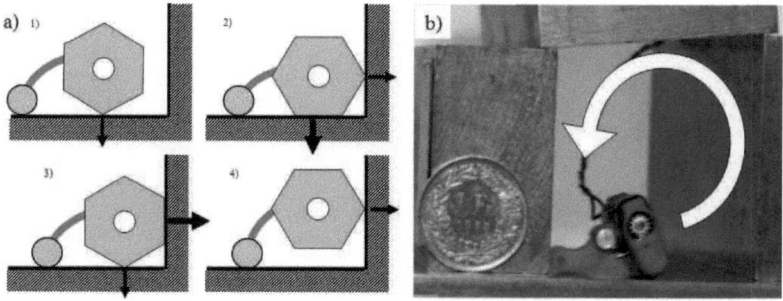

Fig. 5-24: The rotating magnetic cam disc [15]
(a) Basic concept, (b) Successful tests with a low-cost prototype

With these promising basic measurement results, a low-cost prototype was realized to prove the concept in a more illustrative way. As expected, this prototype performed well both in inner transitions, rounded outer transitions and rusty environments. However, its motion is highly inhomogeneous – which leads to strong vibrations that do not only disturb any camera image but are also a significant risk for onboard NDT-sensors or other sensitive components. Mainly for this reason, a full prototype has not been realized yet.

New mechanisms for wheeled robots have been developed (see 4.2.1.3) as well – the passive double front-arm, the Tri-Wedge and the wheel-parallel-to-the-wheel (WpW). For testing all of them with reasonable effort, a modular test-prototype was derived from the previous prototype (see Fig. 5-25-a). During these tests, the WpW (= wheel parallel to the wheel) resulted the most promising. Its mechanical complexity is the lowest among all 3 mechanisms and it also works on steps and other combined obstacles. A cut view of its basic design and the sequence of passing an inner transition are represented in Fig. 5-25-b and c. More details about the other two tested mechanisms and the main difference between the WpW-technology and the dual-magnetic wheel by Osaka Gas [40] can be found in the chapter about obstacle-passing-mechanisms (4.2.1.3 and 4.2.1.4)

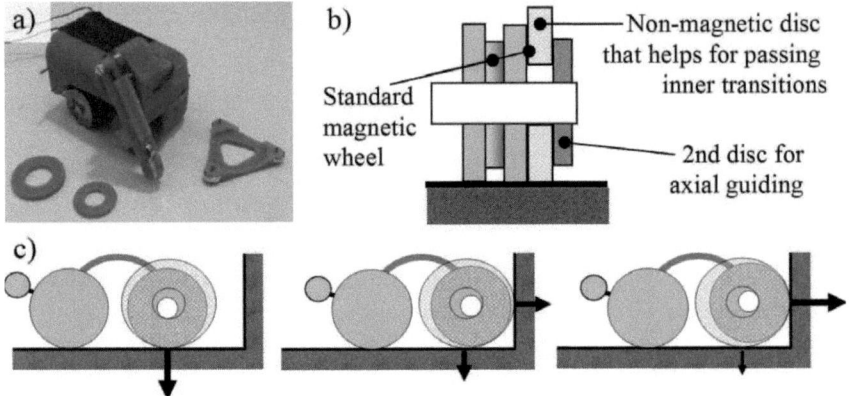

Fig. 5-25: Development of the Wheel-parallel-to-the-wheel (a) Modular prototype for testing three different mechanisms, (b) Schematic cut-view of the WpW-concept, (c) Sequence of passing an inner transition

Soon after its development, we realized that the WpW-technology does not only bring advantages for a very small robot that moves in the generator housings, but is also very promising for bigger robots at the size of the MagneBike – especially if it is combined with additional active mechanisms for edges. This idea finally led to the concept of the MagneBike PCAE (5.2.3.4) – which allows for using this technology in the highly valuable steam-chest application.

5.3.3.3 The Micro-Tripod WpW

A similar approach as in the MagneBike PCAE – combining the WpW-technology with an active structure extension for outer transitions – was then also realized in a robot that is small enough for moving in generator housings: The Micro-Tripod WpW [14]. For keeping the number of actuators as low as possible, the joint in the active structure extension is coupled with the camera movement for looking up or down. By using this active structure extension in outer transitions, the normal force on the main wheels can be increased significantly when it is placed on a sharp edge (Fig. 5-26-b-3). This force increase allows for enough traction even in the worst case. Both motion sequences for inner and outer transitions are represented in Fig. 5-26.

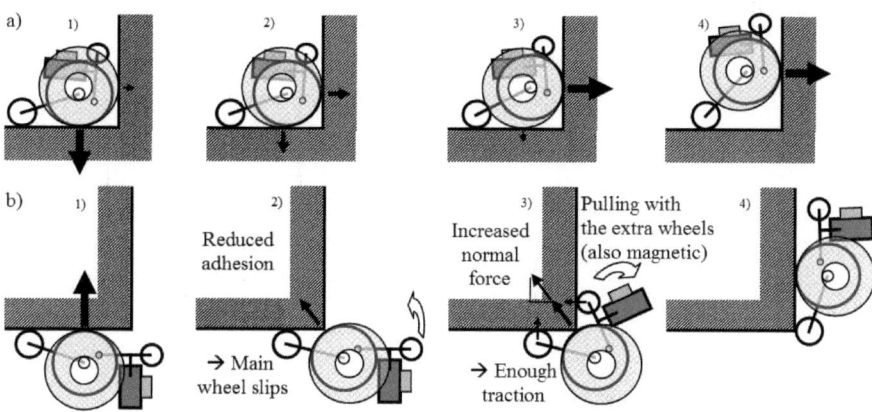

Fig. 5-26: Motion sequences of the Micro-Tripod WpW [14]
(a) Inner transition using the WpW-technology,
(b) Outer transition, where the active edge-extension allows for a significant increase of the traction force on sharp edges

The detailed design has been realized in the student project of Ursin Hutter [14], with the main components represented in Fig. 5-27. In the context of this project, several experiments with different vehicle configurations and in different test environments have been performed as well: WpW-rims with different diameters, different values for the distance between main and back wheels; and configurations with motorized back wheels or rubber on the wheels instead of the WpW.

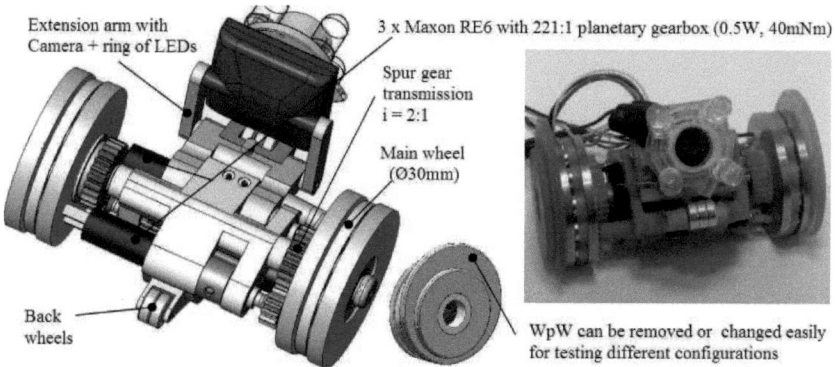

Fig. 5-27: Detailed mechanical design of the Micro-Tripod WpW [14]

While the detailed description of all these experiments can be found in the student report [14], the most important results coming from these experiments can be summarized as follows:

- Concerning the diameter of the WpW, the lowest torque could be achieved with outer rims that are approximately 20% bigger than the main wheels. For choosing the best value for the distance between main and back wheels (L), a value of approximately 1.5 times the main-wheel radius showed the best results both on inner and outer transitions. While longer vehicles need more torque for passing inner transitions, shorter ones easier lose contact on the back wheel during outer transitions when descending from a flat summit-plateau into a vertical wall.

- Within the passive mechanisms for helping in inner transitions which have been tested in the context of this student project, the WpW is the only one that does not cause negative effects on outer transitions. As it can be seen in Fig. 5-28-a-1, the configuration without any mechanism implemented was not able to pass corners. Approaching this limitation by covering the wheels with a thin layer of rubber caused the robot to fall down on all outer transitions due to the significantly reduced adhesion force (Fig. 5-28-a-2). Increasing the traction by powering the back wheels also helps in inner transitions and does not disturb on outer transitions from a vertical wall to a top plateau. However, in outer transitions from ceiling to wall the robot gets pushed off the surface by its back wheels which makes it fall down (Fig. 5-28-a-3). Only the robot with the WpW-configuration was able to pass both inner and outer transitions without any problems (Fig. 5-28-a-4).

- For passing outer transitions, the robot was even able to pass the worst-case inclination from ceiling to wall without using the active extension (Fig. 5-28-b). However, when an additional test-payload of 30g was hung on the robot, the wheels started to slip. By using the active edge-extension, the robot could then pass even with the payload attached.

What can be concluded from these tests is the following (Fig. 5-28).

   a) A robot with only front-wheel traction but equipped with the WpW-technology can pass both inner AND outer transitions passively.

   b) Its payload capability in environments with sharp edges can be further increased by implementing an active edge- extension-arm.

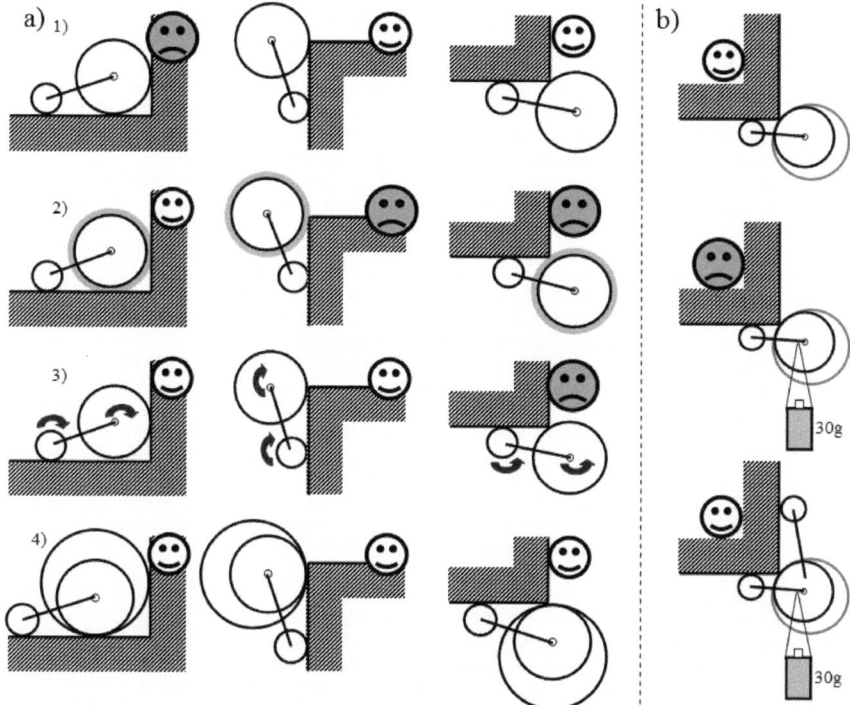

Fig. 5-28: Experiments with the Micro-Tripod WpW in different configurations [14] (a) Mechanisms helping in corners and their influences on the capability to pass edges: (1) Nothing, (2) Thin rubber cover on the wheel, (3) Traction also on the back wheels, (4) WpW
(b) Observation that already the very simple structure can pass edges, but the payload can get further increased by using the active edge extension

Within these test results, the observation that the robot could pass outer transition from ceiling to wall without using the edge-extension is the most interesting – as this ability had not been predicted according to the previously used calculation model (Fig. 5-29-a). According to this model, the robot would need an infinite traction force at the worst case – which is of course not possible and in reality causes the wheel to slip already before this case is reached.

The observation that the robot is indeed able to pass can be explained by taking into account the edge-pushing-force (see Fig. 4-5) which results from the asymmetric distribution of the adhesion force when the wheel is on the edge. Measurements with the wheel of this prototype (15mm radius, 20N adhesion force) led to a maximum torque of approximately 30mNm – which corresponds to 10% of the magnetic adhesion force multiplied with the wheel radius. This torque, which helps the wheel turning into the direction pointing away from the edge, allows the robot to pass. The basic influence of this effect on the robot is shown in Fig. 5-29-a, while more detailed calculations for the robot prototype can be found in the student report of Ursin Hutter [14].

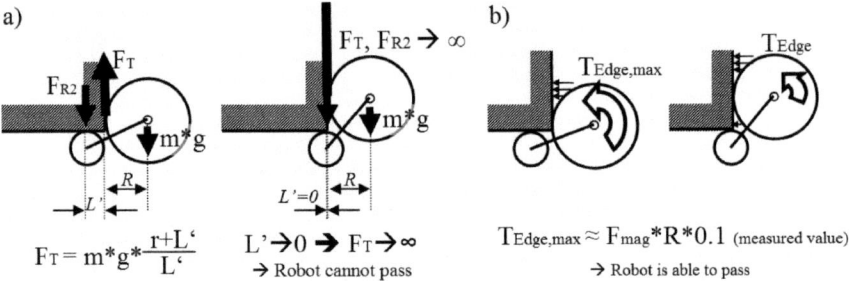

Fig. 5-29: Calculation models for edges [14] (a) Previous model which predicted that the robot would not pass, (b) Extension for the model which takes into account the edge-pushing-force (see Fig. 4-5)

5.3.3.4 Externally developed prototypes with similar properties

Besides the robots developed in our group at ETH, at the same time also our partners at EPFL and other research teams realized robots at similar size and with passive mechanisms for inner transitions. Some of them show interesting approaches and are presented in a very professional way. However, to our best knowledge none of these robots fulfills all mandatory specifications in generator housings. The main limitation for most oft these prototypes is the missing ability to pass outer transitions both from a ceiling to a wall AND from a wall to a top ground. At least, in none of the analyzed publications and videos, the robots are shown in both obstacles.

In Fig. 5-30, a selection of externally developed robots with high similarity to the ETH-prototypes for generator housings is presented and compared against the most difficult mandatory specifications in this scenario – showing both their main limitations there and additional functions/features that could potentially bring advantages in other types of applications.

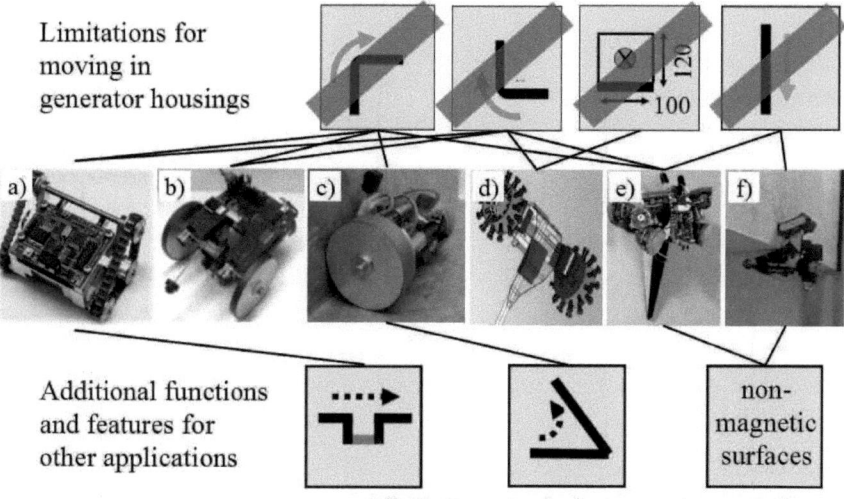

Fig. 5-30: Robots developed by other research teams, their main limitations in generator housings and additional functions/features they offer (a) Tripillar [39], (b) Robot with turning adhesion zone and stick [44], (c) CyMag [45], (d) Gecko [46], (e) WaalBot [84], (f) Gel-type sticky mobile inspector [85]

5.3.4 Innovation, outlook and lessons learned

5.3.4.1 Innovations

Concerning the number of new concepts, this scenario can be seen as the one with the highest activity: 6 innovative concepts by our team (structure with passive extra wheels, rotating magnetic cam disc, passive-front-arm, tri-wedge, wheel-parallel-to-the-wheel, edge-extension coupled with camera movement), at least 3 by our partners at EPFL, and several others by external research teams.

The wheel-parallel-to-the-wheel-technology led to the development of the MagneBike PCAE ([9] and chapter 5.2.3.4) which allowed for patenting it [10] in the context of the project on steam chests inspection with a high business value. Also the basic conceptual idea of the Cy-Mag [45] – developed by our partners at EPFL – has been patented by ALSTOM.

5.3.4.2 Outlook

To our best knowledge, the business value of this application was never as high as the ones in steam-chests or generator-air-gaps (5.4.1). For this reason, the industrialization of any of the prototypes has not started yet.

5.3.4.3 Conclusion and lessons learned

Even if not offering a high business value, this project created a lot of important results: Most of the passive obstacle-passing mechanisms for inner transitions (see chapter 4.2.1) have been developed, analyzed and compared in this context; and also the calculation models for inner- and outer transitions could be improved and extended. These new mechanisms and models could be very helpful in future developments on compact inspection robots in similar environments.

Furthermore, the successful application of the wheel-parallel-to-the-wheel-technology into the MagneBike-project can be seen as an example of successful technology transfer between similar projects. As it will be shown more detailed at the end of this chapter, such technology transfer usually goes from difficult to easier projects and not the other way round.

For this reason, typical "basic-research"-projects like this one (technically challenging but relatively low business importance) should always be considered with a significant part of the available resources. Even if they sometimes do not pay back immediately, their results can form a solid base for long-term success in "applied-research"-projects, which are usually less challenging from the technical point of view, but offer a high business importance (e.g. steam chest).

5.4 Successful projects with other technical challenges

Besides the three main projects, also other projects on compact climbing robots for inspection tasks have been realized in the context of this work – generator air gap, helicopter dock, foldable turbine crawler, and cable-crawler. The technical challenges were different from those in the main projects, but the scientific success was remarkable: All three conference papers that were published before May 2010 have been invited for later-on journal-publications. The paper about the foldable micro-crawler for turbine-inspection [24] even won the Industrial Robot Innovation Award 2009. High economic success could be achieved as well in one of these projects: The last generation of the generator air gap crawler led to a patent [21] and an industrial prototype with high business value.

5.4.1 Generator air gap

The inspection of generator stators with the rotor still installed is one of the most traditional applications for compact climbing robots in power plant inspection – with the first patents dating back more than 20 years [38]. Compared to the other scenarios, the obstacles in this environment are less difficult – only gaps/holes that can be passed with a row of several wheels or caterpillars (see Fig. 5-31-c). However, the height restriction is very challenging for the robot design, with approximately 12.5mm (0.5 inch) in most generators; and down to 9mm in the narrowest ones (see Fig. 5-31-b).

For this reason, compact mobile robots from other teams already existed at the start of this project – the Siemens FastGen (Fig. 5-32-a-1, [37]) running on magnetic tracks; and the GE-Magic (Fig. 5-32-a-3, [58]) that spreads between stator and rotor. Both robots achieve a maximum height of 12.5mm. At ALSTOM, the only available systems were the DIRIS Flex with more than 50mm of height (need to remove the rotor, see Fig. 5-32-a-1, [36]) and a system similar to a cable-car which is very time-consuming to install (Fig. 5-32-a-4, [92]). The goal of this project was to develop a new robot with similar properties as the competitor products but somehow better. This "better" mainly meant to reduce the height down to even 9mm, and to additionally achieve mobility in axial and circumferential direction. Ideally, the circumferential paths should even be realized without turning the vehicle – as it is the case in the GE Magic [58].

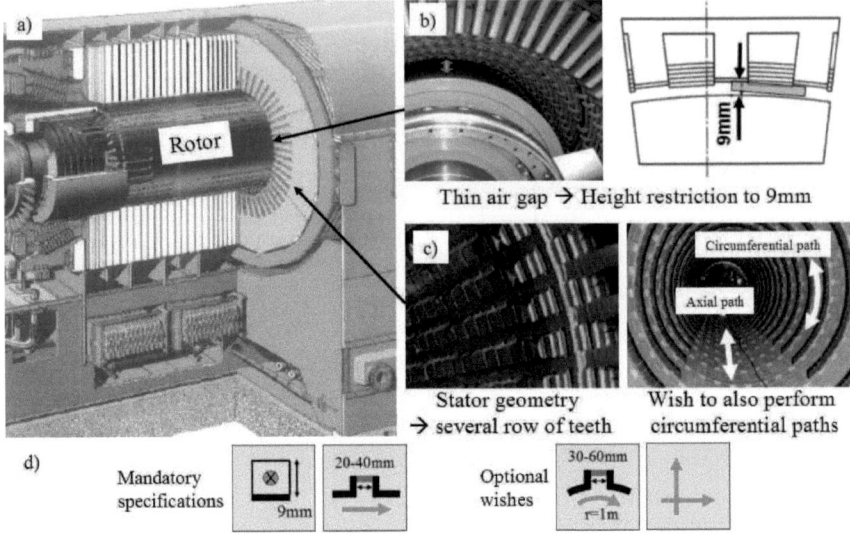

Fig. 5-31: Requirements for the generator air gap scenario (a) CAD-model of a generator, (b) Height restriction at the air gap, (c) Stator geometry with several rows of teeth, (d) Specifications for a mobile robot

A basic concept for such a robot was quickly generated and approved. It uses a state-of-the-art crawler on magnetic wheels for the axial paths (Fig. 4-26-a-3); and an extension for inchworm-movements in circumferential direction (see Fig. 4-29-b). After some preliminary analysis and several iterations for the mechanical design, the final prototype was completed in the semester work of Johannes Berkenhoff [17] and tested in an artificial environment at our lab (Fig. 5-32-b-1). However, making it reliable enough for also moving on real stators, and robust enough for industrial use seemed out of scope – mainly caused by the high mechanical complexity.

For this reason, a simpler design was developed as well – the G1-double-flexible (Fig. 5-32-b-2). It uses two wide units that span over several rows of teeth and differential drive for steering (Fig. 4-28-b) – similar to the GE Magic. In contrast to this competitor product, our robot however generates its adhesion by using several magnets on flexible shafts – which allows for fewer parts and lower height than mechanically spreading (like in the GE Magic). This technology was then patented by ALSTOM [21].

For first tests and publications, simplified versions of both designs have been realized as well (Fig. 5-32-c) – the axial drive unit of the firsts concept [16] and a preliminary design of the final concept with flexibility in only one direction [19] – using one motor, a flexible shaft and Mod0.2 worm-gears between the flexible main shaft and the magnetic wheels. The paper about the axial drive unit [16] received excellent review results, which allowed for publishing the full version with circumferential mobility in Transactions on Industrial Electronics [18].

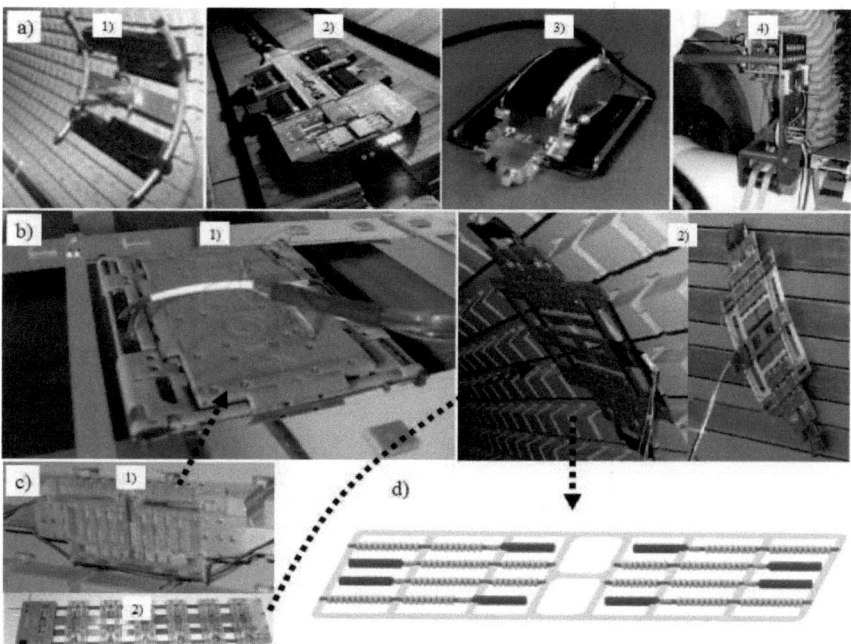

Fig. 5-32: Robot prototypes developed for moving on generator stators (a) External developments done before the project start (1) DIRIS Flex [xx], (2) Siemens FastGen [37], (3) GE Magic [58], (4) DIRIS Cable car (b) Main prototypes in this project (1) Robot with inchworm motion in circumferential direction [14], (2) Robot with flexible magnetic shafts [21], (c) Preliminary prototypes (1) Axial drive unit [14], (2) First test prototype with flexible shafts [19, 20], (d) Concept for the final industrial version

A more robust version of the last design with flexible magnetic shafts is currently under development at ALSTOM Inspection Robotics and in collaboration with our team (Fig. 5-32-d). This last generation is planned to be industrialized in a small series of approximately 5 units – similar to the MagneBike-prototypes.

5.4.2 Lightweight helicopter dock

The knowledge and the technologies which were generated in the project on the robots for the generator air gap could also be reused successfully in another project on boiler inspection [23]. In this project, the basic idea is to use a small unmanned helicopter (quad-rotor, ~500mm width and 1-2kg mass) for carrying the inspection sensors and docking on the ferromagnetic walls of the boiler (Fig. 5-33-c). This approach was chosen, as boiler walls usually are too dirty for moving with magnetic wheeled robots. Our part of the project consisted in the development of a robust and lightweight device that allows for a variable adhesion force between 15N and almost zero on rusty dirt-covered walls – building on the results from a pre-study in the semester thesis of Tobias Hänggi [22].

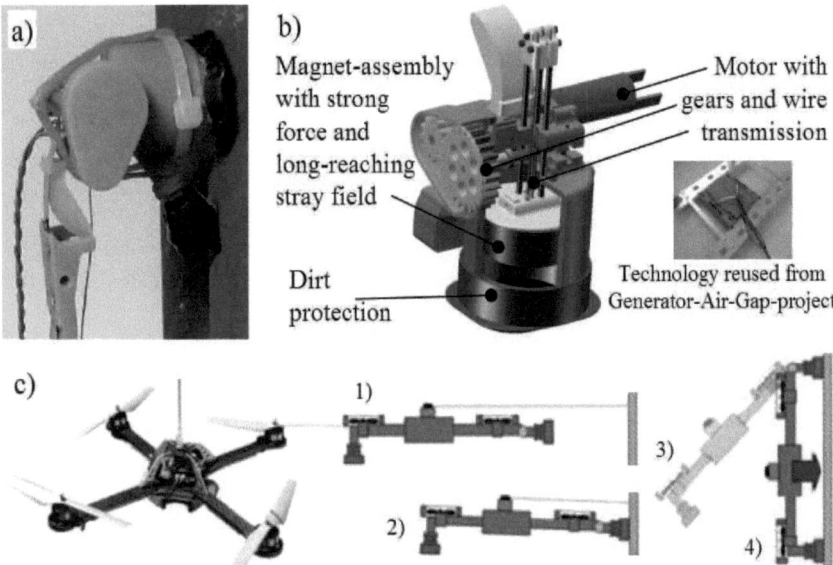

Fig. 5-33: Lightweight helicopter dock [23] (a) Photo of the prototype, (b) CAD-model showing the core components, (c) Docking procedure (taken from the student project of Tobias Hänggi [22])

For solving this design challenge, the experience in small wire-transmissions which was achieved during the design of the generator air gap crawler could be successfully reused. With this knowledge, a functional prototype could be realized in less than one week of work. It uses the same motors as most of our other climbing robots at very small size (Maxon RE6 with 221:1 gearbox, 40mNm), a pulley-transmission of 2:1 on an Ø1mm-shaft, and an additional stage of spur-gears with a reduction of 2:1. With these components, it is able to detach a magnet assembly of up to 200N adhesion force, while the mass of the device is less than 20g. For the adhesion, standard ring-magnets from a previous project ($F_{adh} \approx 20N$, m=10g) are used. Even if these magnets are not the optimum yet, the required specifications could already be fulfilled successfully. By using stronger magnet assemblies, the ratio between mass and maximum adhesion force will be improved even further in the next generation.

5.4.3 Micro-crawlers for boiler tubes and steam turbines

For boiler tubes, we performed a pre-study on simple robots at very small sizes in the range of around Ø10-25mm and proposed a new design for a locomotion unit with magnetic wheels and worm gear transmission. Before handing the project to our partners at EPFL, we could already realize and successfully test a prototype of such a drive unit (Fig. 5-34-a). More details about the final prototype of the tube-crawler can be found in the semester thesis of O. Nguyen [34].

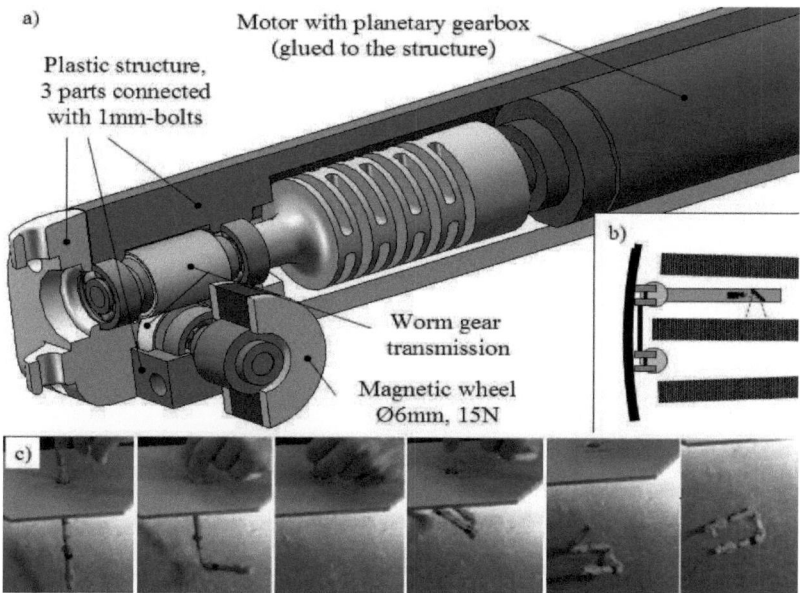

Fig. 5-34: Foldable micro-crawler for turbine inspection [24]
(a) CAD-model of the drive unit, (b) Analysis of its potential application in turbine inspection, (c) Sequence of the folding-mechanism

Additionally, we also realized a concept of how to couple two of these units to a very small tripod with the ability to pass through a very small entrance hole and then unfold afterwards (see Fig. 5-34-c). This concept was successfully realized in a prototype and analyzed for a potential application in turbine inspection (see Fig. 5-34-b). Even if not chosen for industrialization at ALSTOM yet, this prototype is the only climbing robot with 2D-mobility that still fits through Ø15mm entrance holes. The publication about this robot which was presented at CLAWAR09 won the Industrial Robot Innovation Award 2009 [24, 25].

5.4.4 Cable crawler for power line inspection

In the context of the focus project 2007, 6 students at ASL developed a robot for power line inspection – with the goal to look for trees that grow into the cables. Thanks to its unique combination of rollers and springs, this prototype cannot only climb on slightly inclined cables, but is also able to passively roll over mast tops (Fig. 5-35-b) without using actively powered joints in the structure. This ability was achieved for the first time, while comparable robots (e.g. the Line-Scout [63]) need a significantly higher system complexity. During the design phase, the experience gained in the other projects on compact climbing robots was an important help.

Fig. 5-35: Cable-crawler for power line inspection [26]
(a) Mast with power lines, (b) Prototype passing a mast top

The publication about this robot [26] achieved very good review results as well. For this reason, it was also nominated for the Industrial Robot Innovation Award 2009 and invited for later-on journal publication [27].

5.4.5 Innovation, outlook and lessons learned

5.4.5.1 Innovations

Even if the technical challenges in these projects were different from the ones encountered in the main projects, many innovative solutions have been generated. The one with the highest industrial relevance is the "flexible magnetic shaft", which was developed for the generator air gap project. This technology allows the robot for spanning over gaps both in axial and circumferential direction while still keeping the necessary flexibility that is required for adapting to the large curvature of the stator. Given the high economic value of its application, this technology was patented by ALSTOM [21]. The innovations in the other projects are maybe of less economic significance, but also impressive results.

5.4.5.2 Outlook

The last generation of the generator air gap crawler with flexible magnetic shafts will be industrialized, produced in a small series and very likely become a useful tool for speeding up the inspection of generator stators. The other projects are not planned for industrialization yet.

5.4.5.3 Conclusion and lessons learned

What can be observed at first glance when analyzing the project on generator airgaps is very similar to the steam chest project: In applications with high economic relevance, the likelihood for success is much higher than in projects which only offer technical challenges. For this reason, a strong focus should of course be put on these projects if there is a choice on several ones.

Another important observation is the way how ideas, concepts and technologies were transferred between different projects. As it can a be seen in Fig. 5-35, this transfer usually went from the technically challenging projects to the less challenging ones – not only bringing innovative solutions to the "applied-research-projects" with high economic value (steam-chest and generator-air-gap) but also allowing for impressive additional results in less relevant side-projects at relatively low effort (e.g. helicopter dock: less than 2 weeks for the implementation). For this reason, it can once more be recommended to create and well document a wide knowledge-base of concepts and to always think of potential applications beyond their originally planned application. If possible, a large number of projects should be combined, ideally with examples from all following groups:

5.4. Successful projects with other technical challenges

- Basic-research-projects = very difficult challenges → Create innovation.
- Applied-research-projects = high business-relevance → Satisfy industry-partner.
- Side-projects → Impressive additional results at relatively low effort.

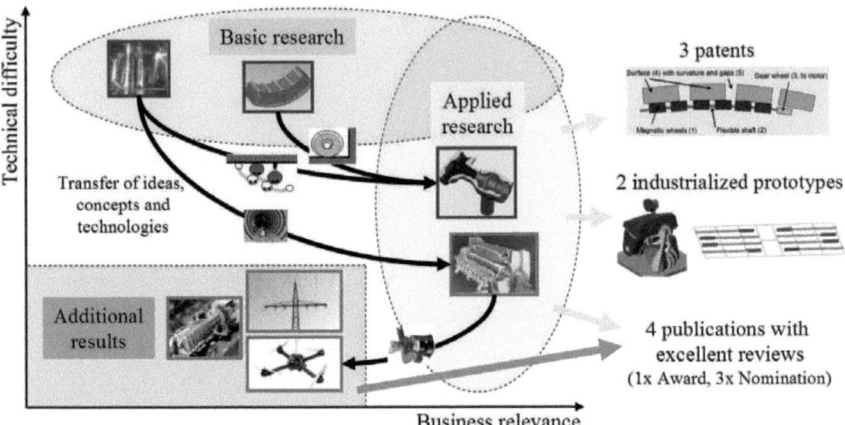

Fig. 5-36: Portfolio representation of the most important research-projects in this thesis, classified according to their business-relevance and technical difficulty, structured into 3 groups, pointing towards the transfer of ideas between projects and concluding with the most impressive results

Thanks to the combination of several different projects in this thesis, the large number of impressive results could get achieved: 3 patents, 2 prototypes that get industrialized, 1 awarded and 3 award-nominated publications; and several other impressive prototypes (>10) and scientific publications (16).

When analyzing our most successful publications about innovative robot design, one remarkable observation can be concluded as well: In all these papers, the main focus is set on well describing the application and the advantages for the customer in comparison to previous designs, while the description of calculation models or other relatively complex theories is kept as short as possible.

Concerning the patented innovations, all of them are relatively simple mechanisms that allow for significantly reducing the mechanical complexity of the robot compared to previous designs – resulting in higher robustness and smaller size. Innovations with their main focus on a mobility-increase were regarded less relevant from the industrial point of view and for this reason not patented.

6 Summary and core value of this work

In this thesis, a comprehensive overlook has been provided on the most recent types of compact climbing robots with high mobility on specific obstacles; and on their application in power plant environments.

Summary of the work and its main contributions

After explaining the most important challenges in these applications, the basic types of climbing robots have been classified according to these needs and compared against each other. The different mechanisms for passing specific obstacles with magnetic wheeled robots were described as well, classified and compared in more detail. For providing a relatively complete overview on most robots and their performance in typical environments, a comprehensive overview matrix with more than 30 prototypes and 45 environment challenges has been established. Seven typical applications with difficult challenges are described in the case studies at the end of this work, three of them in more detail. These case-studies include the economic relevance of each application, the explanation of the most difficult challenges for robot mobility and size, the description of the most relevant innovations and robot prototypes, and their comparison against each other. At the end of each case study, an overlook on the core innovations, an outlook to the future of the project and some lessons learned for similar projects were concluded.

Within these innovations, four of them are of important economic significance for ALSTOM and form the basis of at least two industrial products that will make power plant inspection faster, safer and more attractive. All other innovative mechanisms that are described in this work are not patented and can be re-used by any other research team working on future climbing robots. Together with their structured analysis, classification and comparison (chapter 3+4), these mechanisms form a solid knowledge-base for future projects in the field of compact climbing robots – which will not only bring advantages in the field of power plant inspection, but also in similar tasks such as search-and-rescue, surveillance or anti-terrorist-duty. Additionally, the recommendations concluded at the end of the case-studies can be an important help for future research teams in this field. v

Economic value

In the context of this project, two basic innovations and one pair of innovations (in the context of the MagneBike PCAE) have been identified as economically significant for ALSTOM and filed as patents:

- 2007: The rotary lifter for the first generation of the MagneBike.

- 2010: The flexible magnetic shaft for spanning over small gaps in generator stators, while still being able for flexibly adapting to large curvatures.

- 2010: The Wheel-parallel-to-the-wheel-technology (WpW) which allows for passively rolling through inner transitions at minimal complexity (main claim); combined with the robot structure of the MagneBike PCAE that allows for extremely high mobility at still reasonable complexity (sub-claims)

Based on these patents, two groups of products have been successfully industrialized – the MagneBike-fleet and the Generator Air-Gap-Crawler.

- The first generation of the MagneBike has already been produced in a small series of 5 units which are basically used for two different purposes: Direct application at ALSTOM; and future research on climbing robots that is not directly related to the improvement of the robot mobility any more – 3D-SLAM, advanced inspection methods and multi-robot-coverage.

- The final design of the MagneBike PCAE wheel units – for upgrading the "standard" MagneBikes to robots with higher mobility for more difficult applications – was still under development at the submission-date of this thesis. Assuming positive test results, these technologies should be industrialized as well.

- Also the final industrial version of the Generator Air-Gap-Crawler with flexible magnetic shafts was still under development at the submission-date of this thesis – however already at the main responsibility of Alstom-Inspection-Robotics.

- The industrialization of the MicroTripod WpW has not been planned detailed yet, but could also be imagined.

With all these successful new robots, the inspection services provided by ALSTOM will very likely become faster, better and more attractive than those offered by their competitors. It can be expected that these advantages will soon pay back the monetary investment in this work, and will help to maintain their leadership in the inspection business – which is crucial for several high-tech-workplaces in Switzerland and the European Union.

6. Summary and core value of this work

Scientific value 1: Other innovations

Not only our industry-partners, but also the scientific community can profit from the findings in this thesis, as many innovations have not been patented and thus can be used by everybody. Even if some of them may already look outdated compared to the newest concepts, the ones listed below could still be promising in specific applications.

- Two compact climbing robots in mother-child-structure do not only allow for a very fast and lightweight child robot, but can also avoid turning the robot in narrow spaces.

- Similar to the WpW-technology, also the rotating magnetic cam-disc and other mechanisms in the same group allow for passing inner transitions passively – with their specific advantages and disadvantages explained in chapter 4.2.1.

- Pulling magnets by using wire-transmission with extra reduction allows for very lightweight magnetic feet of variable force that cannot only be used in legged climbing robots but also in docking-modules for micro-helicopters.

- Foldable structures in climbing robots allow for accessing environments with narrow entrance holes and geometries that cannot be scanned with bore-scopes.

Scientific value 2: Overview, classification and evaluation tools

Besides the usefulness of these innovations, also the structured classification and comparison of climbing robots can become an important help in future research projects on compact climbing robots – with the central contributions formed by the prototype-comparison-matrix (appendix 1), the theoretical calculation model for estimating the risk of slipping in inner transitions (most of our papers, especially the one about the first test prototype [3]) and the detailed description and comparison of more than 30 recently developed obstacle-passing-mechanisms in compact climbing robots (chapter 3).

Scientific value 3: Case-studies and recommendations

The recommendations concluded in the case-studies can be very useful as well for future research teams – not only if they work in the field of compact climbing robots. For this reason, the most important ones are again summarized.

- Strong interactions between similar projects can increase the innovativeness significantly – especially when previously rejected ideas and concepts are well documented so that they can be reused in the next project (see Fig. 5-36).

- For being successful in applied research, a strong focus should be put on projects that offer high business relevance (e.g. steam chest inspection).

- However, other projects should also be considered with a reasonable part of the available resources. Ideally, several projects can be combined: Basic-research-projects for creating a solid knowledge-base which is crucial for long-term-success, applied-research-projects for satisfying the industry-partner, and side-projects for generating impressive additional results at relatively low effort.

- When analyzing the most successful innovations (= the 3 concepts patented and industrialized by ALSTOM), it can be observed that all of them aim for a simplification compared to previous systems and are relatively easy to understand even by non-experts.

- Concerning the most successful publications about innovative robot design (1 awarded, 3 nominated), all of them are relatively short and focus on the comprehensible description and the advantages of the core innovation rather than on complex theoretical models.

Appendix 1: Comparison Matrix

How to read it:

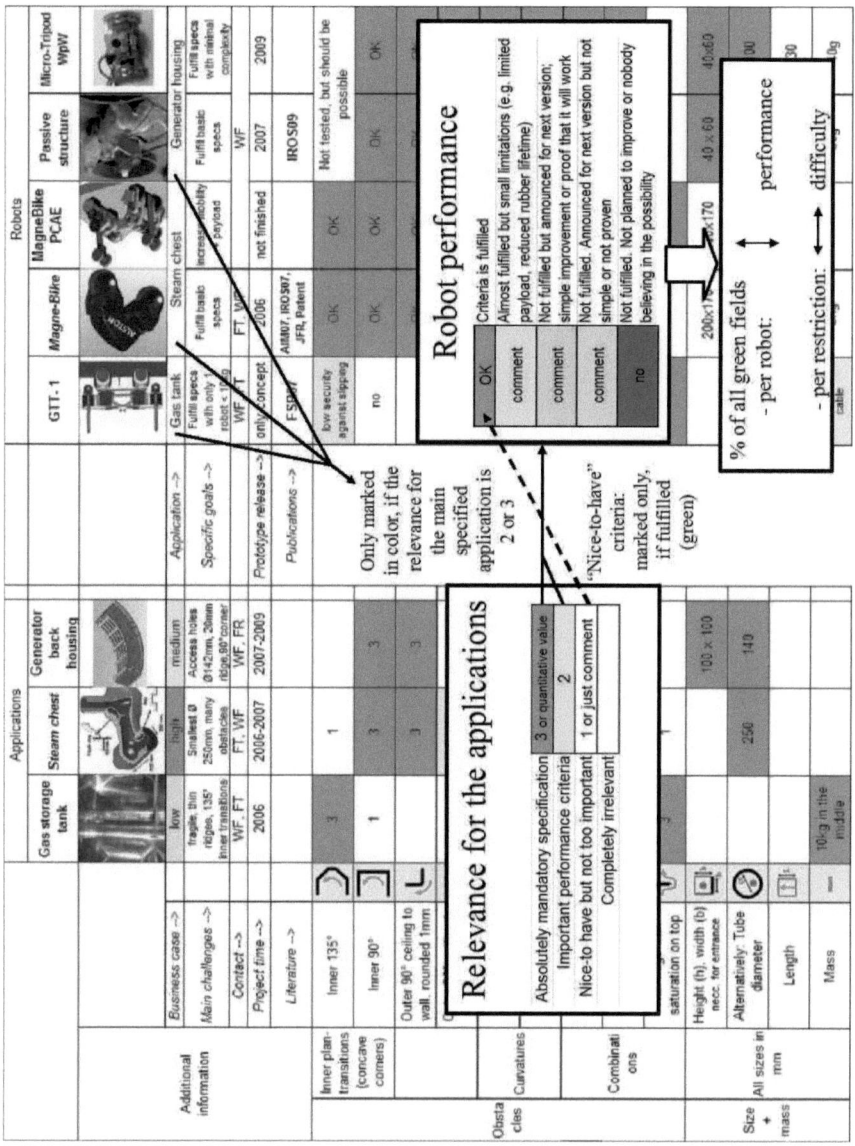

164

Appendix 1: Comparison Matrix

				Robots for thin + fragile surfaces			Roll-legged robots for complex-shaped ferromagnetic environments								non-magnetic alternatives		
Applications + importance of criteria	Gas storage tank	Steam chest	Generator back housing	GTT-1	GTT-2 mother	GTT-2 child	Magne-Bike	Magne-Bike PCAE	Passive structure	Modular test prototype (Wyss)	Magnetic cam-disc	Micro-Tripod VipW	TriPillar	CyMag	SpokeWheel	Gecko (magnetic)	WaalBot	Gel-type Sticky mobile inspector	ALICIA VTX
Hazards								Only relevant for gas storage tanks										
Thin → saturation b=0.7	3		3		not relevant any more	OK												
Paint or ceramics (thin non-magn. layer)				OK			OK	OK	reduced payload, but OK	reduced payload, but OK	reduced payload, but OK	reduced payload, but OK	cannot pass small tubes	cannot pass small tubes	reduced payload, but OK	reduced payload, but OK	OK	OK
Paint drops or weld beams		1 (thin layers <0.5mm)	1				Not tested as mainly relevant for generator air gaps. Generally, bigger wheels (e.g. MagneBike 60mm) perform better.											
Rust or dirt	Remove, as very similar to the tube with.			reduced lifetime of rubber			reduced lifetime of rubber	reduced lifetime of rubber	reduced lifetime of rubber	reduced lifetime of rubber	OK	reduced lifetime of rubber	reduced lifetime of rubber	OK	reduced lifetime of rubber	OK	OK	OK
Short non-magnetic zones < 50mm				no	no	no	yes	OK	no	no	no	no	no	no	no	no	OK	OK
Non-magnetic material				no	no	no	no	no	no	no	no	no	no	no	no	no	OK	OK
Underwater 0.5 bar				Not relevant for gas storage tank application						Only relevant for boiler tubes								
Underwater 10? bar																		
Size + mass All sizes in mm	Height (h), width (b) necc. for entrance		100 x 100				25?, but not with all sensors still look in paper		41 x 50	60 x 60	26 x 60 (1wheel unit)	35×50	50 x 95	28 x 55	? But too big	b > 100	<100x100	<100x100	b > 100
Alternatively: Tube diameter		250	140						100	140	100	200	cannot pass small tubes	cannot pass small tubes	? But too big	?	?	?	?
Length									40	60	30	30	60	28	?	?	?	?	?
Mass (g)	10g in the middle			20g, but not relevant		1kg	3kg		50g	200g	20g	30g	225g	80g	?	?	?	?	?
Extras	Camera	1	3	3	10g without cable	no	no	yes	planned for future version	without service to mirror it	planned for future version	planned for future version	planned for future version	planned for future version	planned for future version	planned for future version	planned for future version	yes	yes
Battery + remote control		1	1	no	no	no	planned for future version												
Other sensors	for navigation see PhD FT	vibration sensor		planned for future version	not relevant any more	planned for future version	Hokuyo, Inclinometer		Preliminary prototypes to test the locomotion concepts, but mostly enough payload to carry additional sensors										
Mechanical guiding elements	helium sniffer possible				yes		yes												
Specific robot properties	N° of active DOF				10	1	5 (steam gauges)		2	2	2	3	2	2	3	2	2	27	2 + vortex airscrew
N° of sensors necc. for mobility	Low complexity is always desired but be never clearly specified			4 (touch sensors)	4 (touch sensors)	0	5 (strain gauges)		0	0	0	0	2 (IR proximity)	0	?	0	0	?	?
N° of wheels (powered+passive)				4+4	8	2+2	2+4	4+2	4+4	4+2	2+1 (sliding contact area)	2+2	6 (2 triangular caterpillars)	2	16	2+2 sliding tail	2 wheels + sliding tail	2 wheels + sliding tail	4 (+ vortex airscrew)
Quantitative criteria	Payload	sensor 1kg + long cables	500g	50g	1kg	9kg	1kg	500g	>500g	20g	not tested	30g	no space	no space	not tested	?	?	no	?
Speed	20m/min	2.5mm/min		too slow	nor relevant any more	fast enough	2.5mm/min	> 2.5mm/min											
Qualitative criteria	Few wheel slip for accurate odometry			OK, but also not necessary			OK	OK	no	no	no	OK	OK	no	no	OK	no	no	no
Continuous movement (for some sensors)			2				OK	OK	OK	OK	OK	OK	OK	OK	OK	OK	?	OK	OK
Static stability (for vibration measurement)			3				OK	OK	OK	OK	OK	OK	OK	no	OK	OK	?	OK	OK

	Magnetic-AirGap-Crawlers				no magnets	Other micro-robots: only 2D-climbing			Tube crawlers			Robots that are not packaged		Applications + importance of criteria				Additional information
G1 basic	G1 circumf. inchworm	G1 flexible	G1 double flexible	GT double flexible new	Siemens FastGen	GE MAGIC	G1 double flex simplified	NanoPod	Foldable micro crawler	Tube crawler 1st version	Tubulo (inchworm tube crawler)	Flying robot with dock	MIT Inchworm or other Bipeds	Generator Air Gap	Boiler tubes	Boiler	Other appl. for micro-robots	
									Gas turbines	Boiler tubes	Boiler tubes	Boiler	Bipeds (+Bodies?)	← Application				
Simple design for small paths of only 9mm height	Omnidirection al mobility using inchworm locomotion for circumf. paths	Better mobility on paint drops, adaptation to curved surfaces	Simple design with 2D-mobility for stators and rotors	Robust industrial version	Competitor products		Extremely simple micro-tripod with 10mm height using modules from the G1-double	Micro-tripod and magnetic wheeled tube-crawler units for underwater use 0.5 bar	Test very small wheeled tube-crawler units and foldable design for tripods	Move tetherless in clean tubes of small diameter	Dirty tubes, underwater, high pressure (107 bar)	Robust + lightweight magnetic dock for micro-helicopters	Biped robot using electromagnetic feet	Allowed height only 9mm, gaps, on circumf. paths combined with curvature	Small size (25 mm) bends 150mm), rust + water up to 107 bar	Extremely dirty and rusty surface → very difficult for wheeled robots. Sometimes fly + dock	Very hard size restrictions, but only 2D-mobility required. Sometimes dirty and wet	← Specific goals
														high	high	medium	medium	← Business case
							not defined yet	?				CH, WF, FR	MIT (Boston)	WF	FR	CH	IS, WF	← Main challenges
2008	2008	2008	2009	planned 2011	Siemens	GE	planned 2011	IS 2009	WF 2008	FR, ON 2009	ON 2009	CLAWAR10	ROS 1996	2007-2009	2008-2009	2008-2009	2009-?	← Contact person / Project time
ICM09	T-IE 2010		Patent (on flex. magn. rollers)		?	?			CLAWAR 09	CARPI 09	CARPI 10							← Prototype release / Publications
OK	OK*	OK	OK				OK	OK	OK			OK	OK					Inner 135° ⌒
OK	OK*	OK	OK				OK	OK	OK			OK	OK					Inner 90° ⌂
OK	OK*	OK	OK				OK	OK	OK			OK	OK					Inner 45° △
										Only climbing on slightly curved surfaces without difficult obstacles		OK	OK					Inner 90°, no rubber tire (µ<0.3) ⌐
												OK	OK					Outer 90° wall to top, rounded 9mm ⌐
												OK	OK					Outer 90° ceiling to wall, rounded 1mm ⌐
												OK	OK					Outer 90° wall to top, sharp ⌐
												OK	OK					Outer 90° ceiling to wall, sharp ⌐
												OK	OK					Outer 90° extra sharp with strong saturation ⌐
												does not fit in such tubes	OK					Double-outer 90°, 20mm thick (sharp ridge) ⌐
										Only 1D-mobility for tubes without bends and slight diameter changes		OK	reduced adhesion					Inner curvature r=1.5m
												OK	reduced adhesion					Inner curvature 250mm
												OK	reduced adhesion					Outer curvature 500mm
												OK	only if smaller than feet					Combination inner-outer curvature
												OK	OK					Step 20-50mm
												OK	OK	3				Tripple step
												OK	OK					Thin ridge with saturation on top (surface fit)
OK	OK*	OK	OK	OK	OK	OK						OK	OK	2 (circumf paths)				Normal hole, 30-100mm
no	OK, but cannot turn to circ-path	OK	OK	no	no	OK						OK	OK					Hole 20-40mm, non-magnetic on ground similar, but with curvature. r=min 1m

Appendix 1: Comparison Matrix 167

Appendix 2: Reference list

For helping the reader to quickly find the relevant one out of 113 references, the list is grouped into own developments, external developments and overviews/classifications. Within the own developments, the papers are assigned to the corresponding case studies (see chapter 5). The external developments are structured in first order according to the adhesion principle. In some cases, also a second order is set for the locomotion principle (see chapter 3).

All referred documents can be found in the CD-attachment. Where scientific publications were not available, the referred web-page was last checked and downloaded in December 2009.

Own publications (including patents and student projects)

Gas storage tanks

1. W. Fischer, F. Tache, R. Siegwart, "Magnetic Wall Climbing Robot for Thin Surfaces with Specific Obstacles", Proc. of The 6th International Conference on Field and Service Robotics (FSR), 2007.
2. W. Fischer, F. Tache, R. Siegwart, "Inspection System for Very Thin and Fragile Surfaces, Based on a Pair of Wall Climbing Robots with Magnetic Wheels", Proc. of The IEEE/RSJ International Conference on Intelligent Robots and Systems (IROS), 2007.

Steam-chests

3. W.. Fischer, F. Tache, G. Caprari, R. Siegwart, "Magnetic Wheeled Robot with High Mobility but only 2 DOF to Control", Proc. of The 11th International Conference on Climbing and Walking Robots and the Support Technologies for Mobile Machines (CLAWAR), 2008.
4. M. Morales, W. Fischer, F. Tâche, R. Siegwart, "Tests und Simulation von Robotern mit magnetischen Rädern", Bachelor Thesis at ETH Zürich, Autonomous Systems Lab, SS 2007.
5. F. Tache, W. Fischer, R. Moser, F. Mondada, R. Siegwart, "Adapted Magnetic Wheel Unit for Compact Robots Inspecting Complex Shaped Pipe Structures", Proc. of The IEEE/ASME International Conference on Advanced Intelligent Mechatronics (AIM), 2007.
6. F. Tache, W. Fischer, R. Siegwart, R. Moser, F. Mondada, " Compact Magnetic Wheeled Robot With High Mobility for Inspecting Complex Shaped Pipe Structures", Proc. of The IEEE/RSJ International Conference on Intelligent Robots and Systems (IROS), 2007.
7. F. Tache, W. Fischer, G. Caprari, R. Moser, F. Mondada, R. Siegwart, "Magnebike: A Magnetic Wheeled Robot with High Mobility for Inspecting Complex Shaped Structures", Journal of Field Robotics, Vol. 26-5, May 2009, 453-476.
8. R. Moser, W. Fischer, F. Tache, R. Siegwart, F. Mondada, "Automotive inspection vehicle", Patent EP 2003044A1, Alstom Technology Ltd., 14.06.2007
9. L. Bagutti, W. Fischer, E. Zwicker, R. Siegwart, "Magnetic wheeled climbing robot passing corners and sharp edges", Bachelor thesis at ETH Zürich, Autonomous Systems Lab, SS 2009.
10. Wolfgang Fischer, "Wheel-parallel-to-wheel", Invention disclosure for a potential patent application at ALSTOM, Zürich, 11.12.2008

Appendix 2: Reference list 169

11. Wolfgang Fischer, "Double-tail-structure", Invention disclosure for a potential patent application at ALSTOM, Zürich, 11.12.2008

Generator housings

12. W. Fischer, G. Caprari, R. Siegwart, R. Moser, "Compact Magnetic Wheeled Robot for Inspecting Complex Shaped Structures in Generator Housings and Similar Environments", Proc. of The IEEE/RSJ International Conference on Intelligent Robots and Systems (IROS), October 2009.

13. M. Oeschger, W. Fischer, G. Caprari, R. Siegwart, "Improvement of a compact inspection robot with magnetic wheels", Master thesis at ETH Zürich, Autonomous Systems Lab, SS 2008.

14. U. Hutter, W. Fischer, E. Zwicker, R. Siegwart, "Realization of a new compact magnetic wheeled climbing robot", Bachelor thesis at ETH Zürich, Autonomous Systems Lab, SS 2009.

15. W. Fischer, G. Caprari, R. Siegwart, R. Moser, "Very Compact Climbing Robot rolling on Magnetic Hexagonal Cam-Discs, with High Mobility on Obstacles but Minimal Mechanical Complexity", Proc. of the 41st International Symposium of Robotics (ISR 2010), Munich, 7-6 June 2010.

Air gap Crawler

16. W. Fischer, G. Caprari, R. Siegwart, R. Moser, "Robotic Crawler for Inspecting Generators with Very Narrow Air Gaps", Proc. of The 5th IEEE International Conference on Mechatronics (ICM), 2009.

17. J. Berkenhoff, W. Fischer, G. Caprari, R. Siegwart, "Weiterentwicklung eines kompakten Sonderfahrzeugs für die Inspektion von Generatorspalten", Semester thesis at ETH Zürich, Autonomous Systems Lab, SS 2008.

18. W. Fischer, G. Caprari, R. Siegwart, R. Moser, "Locomotion System for a Mobile Robot on Magnetic Wheels With Both Axial and Circumferential Mobility but only 8mm Height for Generator Inspection With the Rotor Still Installed", Transactions on Industrial Electronics, 2010, accepted but not published yet.

19. A. Hugelshofer, W. Fischer, G. Caprari, R. Siegwart, "Development of a flexible magnetic wheeled drive unit of very low height", Bachelor thesis at ETH Zürich, Autonomous Systems Lab, SS 2008.

20. W. Fischer, G. Caprari, R. Siegwart, R. Moser, "Flexible robotic crawler at very low height for inspecting generators with internal obstacles", planned for publication.

21. Wolfgang Fischer, "Flexible magnetic shaft", Invention disclosure for a potential patent application at ALSTOM, Zürich, 11.12.2008

Other projects (Helicopter-dock, Turbine Inspection, power lines)

22. T. Hänggi, C. Hürzeler, W. Fischer, R. Siegwart, "Design and Implementation of a Lightweight Docking Mechanism for a Miniature Quadrotor Helicopter", Semester thesis at ETH Zürich, Autonomous Systems Lab, WS 2008/2009.

23. W. Fischer, C. Hürzeler, R. Siegwart, "Lightweight magnetic foot for docking unmanned helicopters to steel walls", Proc. of The 13h International Conference on Climbing and Walking Robots and the Support Technologies for Mobile Machines (CLAWAR), 2010.

24. W. Fischer, G. Caprari, R. Siegwart, R. Moser, "Foldable Magnetic Wheeled Climbing Robot for the Inspection of Gas Turbines", Proc. of The 12th International Conference on Climbing and Walking Robots and the Support Technologies for Mobile Machines (CLAWAR), 2009. Winner of the Industrial Robot Innovation Award 2009.

25. W. Fischer, G. Caprari, R. Siegwart, I. Thommen, W. Zesch, R. Moser, " Foldable magnetic wheeled climbing robot for the inspection of gas turbines and similar environments with very narrow access holes", Industrial Robot, Vol. 37, Issue 3, pp. 244-249, 2010.

26. M. Buehringer, J. Berchtold, M. Buechel, C. Dold, M. Buetikofer, M. Feuerstein, W. Fischer, C. Bermes, R. Siegwart, "CableCrawler - Robot for Power Line Inspection", Proc. of The 12th International Conference on Climbing and Walking Robots and the Support Technologies for Mobile Machines (CLAWAR), 2009. Nominated for the Industrial Robot Innovation Award 2009.

27. M. Buehringer, J. Berchtold, M. Buechel, C. Dold, M. Buetikofer, M. Feuerstein, W. Fischer, C. Bermes, R. Siegwart, " Cable-Crawler – Robot for the inspection of high-voltage power lines that can passively roll over mast tops", Industrial Robot, Vol. 37, Issue 3, pp. 256-262, 2010.

External publications on robot design
Climbing robots with magnetic adhesion
Simple designs on wheels

28. W.Guy, "Magnetic wheel", Patent US 3 690 393, Sept. 12, 1972.

29. S. Mondal, A. Brenner, J. Shang, B. Bridge, T. Sattar, "Remote Automated Non-Destructive Testing (NDT) Weld Inspection on Vertical Surfaces", Proc. of the 11th International Conference on Climbing and Walking Robots and the Support Technologies for Mobile Machines (CLAWAR), September 2008.

30. A. Isola, M. Casella, F. Grassia, "VENOM ROBOT - A Climbing Robot for metal wall", Student Competition -Clawar 2004 -MadridCSIC 22-24 September 2004.

31. Jireh Industries Ltd., "Tripod - Magnetic Remote Transport Vehicle", info brochure, http://www.jireh-industries.com/images/stories/AA-002-web.pdf.

32. ALSTOM Inspection Robotics Ltd., "Products - MRS-Series (Mobile Robotic System)", info on website, http://www.inspection-robotics.com

33. R. Moser, C. Udell, A. Montgomery, "Automated Steam Turbine Rotor Disc Inspection", Proc. of The 6th International Conference on Field and Service Robotics (FSR), 2007.

34. O. Nguyen, F. Rochat, P. Schoeneich, F. Mondada, H. Bleuler, "Versatile train-like tube crawler", Semester Thesis at EPFL, Laboratoire de Systèmes Robotiques 1, WS2008/2009.

35. H. Rodriguez, T. Sattar, J. Shang, "Underwater Wall Climbing Robot for Pressure Vessel Inspection", Proc. of the 11th International Conference on Climbing and Walking Robots and the Support Technologies for Mobile Machines (CLAWAR), September 2008.

Appendix 2: Reference list

Robots on tracks/Caterpillars

36. R. Moser, B. Mark, "Automated Robotic Inspection of Large Generator Stators", Proc. of The IEEE/ASME International Conference on Advanced Intelligent Mechatronics (AIM), 2007.
37. D. Schoeler, P.O. Miranda; "Improved Unit Reliability & Availability Through Optimised Predictive Maintenance", Siemens AG, Power Generation; 2003. http://www.energy.siemens.com/hq/pool/hq/energy-topics/pdfs/en/service/3_Improved_Unit_Reliability.pdf
38. G. Dailey, P. Morrison, "Rotor-in-stator examination magnetic carriage and positioning apparatus", Patent US 4 803 563, Westinghouse Electric Corp., 2.9.1987.
39. F. Rochat, P. Schoeneich, O. Truong-Dat Nguyen, F. Mondada "TRIPILLAR: miniature magnetic caterpillar climbing robot with plane transition ability", Proc. of The 12th International Conference on Climbing and Walking Robots and the Support Technologies for Mobile Machines (CLAWAR), 2009.

Mechanisms for inner transitions

40. Y.Kawaguchi, I.Yoshida, H.Kurumatani, T.Kikuta, and Y.Yamada, "Internal pipe inspection robot", in Proc. of the IEEE International Conference on Robotics and Automation (ICRA'95), Nagoya, Japan, May 1995, pp. 857–862.
41. R. Pelrine, E. Edwards, L. Gullman, "Vehicle adapted to freely travel three-dimensionally and up vertical walls by magnetic force and wheel for the vehicle", Patent US 5 220 869, Osaka Gas Company Ltd., June 22, 1993.
42. C. Groux, F. Tâche, F. Mondada, R. Siegwart, "Micro-robot d'inspection à roues magnétique", Semester Thesis at EPFL, Autonomous Systems Lab, SS 2006.
43. A. Prodan, F. Rochat, P. Schoeneich, P. Noirat, F. Mondada, "Spokheel. Autonomous robot with high mobility and magnetic adhesion", Semester Thesis at EPFL, Laboratoire de Systèmes Robotiques 1, SS 2009.
44. X. Wang, F. Rochat, P. Schoeneich, F. Mondada, H. Bleuler, "Climbing robot with magnetic adhesion", Semester Thesis at EPFL, Laboratoire de Systèmes Robotiques 1, SS 2008.
45. B. Lüthi, F. Rochat, P. Schoeneich, P. Noirat, F. Mondada, "Magnetic Wheel - Robot d'inspection miniature", Semester Thesis at EPFL, Laboratoire de Systèmes Robotiques 1, WS2008/2009.
46. A. Shapiro, "Gecko – Climbing robot for complex surfaces (BGN)", Document downloaded from BGN Technology Transfer Company of Ben-Gurion University, http://www.ittn.org.il/technology.php?cat=0&BGN&tech_id=23199
47. R. Erne, F. Tâche, G. Caprari, R. Siegwart, "Flexibles Magnetrad für Roboter in komplex geformten Rohrstrukturen", Student project at Autonomous Systems Lab, ETH Zürich, SS 2009.

Hybrid and legged robots

48. M. Suzuki, T. Yukawa, Y. Satoh, H. Okano, "Mechanisms of Autonomous Pipe-Surface Inspection Robot with Magnetic Elements", 2006 IEEE International Conference on Systems, Man, and Cybernetics, October 8-11, 2006.

49. Kotay, K. and Rus, D. 1996. Navigating 3d steel web structures with an Inchworm robot. In Proc. of the 1996 International Conference on Intelligent Robots and Systems, Osaka, 1996.
50. A. Mazumdar, H. Asada, "Mag-Foot: A Steel Bridge Inspection Robot", In Proc of the 2009 IEEE/RSJ International Conference on Intelligent Robots and Systems, St. Louis USA, October 11-15, 2009.
51. DIEES - Università di Catania – Robotica, "ROBINSPEC - ROBot for INSPECtion", http://www.robotic.diees.unict.it/robots/robinspec/robinspec.htm,
    A detailed scientific publication exists, but is difficult to access: L. Fortuna, A. Gallo, G. Giudice, G. Muscato, "ROBINSPEC: A Mobile Walking Robot for the Semi-Autonomous Inspection of Industrial Plants", in Robotics and Manufacturing: recent trends in research and applications, Vol. 6, ASME PRESS New York (USA), pp. 223-228, March 1996.
52. P. Underwood, F. Kocijan, "Switchable permanent magnetic device"; Patent US 6,707,360, The Aussie Kids Toy Company, March 16, 2004
53. Schmalz GmbH, "Magnetic grippers SGM", Product brochure found on http://www.schmalz.com/np/pg/produkte?hier=155-3886-3891-4009&lng=en
54. J. Berengueres, K. Tadakuma, T. Kamoi, R. Kratz, "Compliant Distributed Magnetic Adhesion Device for Wall Climbing", Proc. of the 2007 IEEE International Conference on Robotics and Automation Roma, Italy, 2007.

Climbing robots with mechanical adhesion

In pipes, tubes or gaps

55. A. Zagler and F. Pfeiffer, "MORITZ a pipe crawler for tube junctions", in Proc. of the 2003 IEEE International Conference on Robotics & Automation (ICRA'03), Taipei, Taiwan, Sept. 2003, pp. 2954–2959.
56. Hydropulsion Ltd., "Vertical Crawler - Vertical Pipe and Duct Inspection System", Info-brochure http://www.hydropulsion.com/robotic-crawler-systems/vertical-crawler/vertical_crawler.pdf
57. K. Suzumori, T. Miyagawa, M. Kimura, and Y. Hasegawa, "Micro inspection robot for 1-in pipes", IEEE/ASME Transactions on Mechatronics, vol. 4, no. 3, pp. 286–292, Sept. 1999.
58. D. Roney, R. Zawoyski, "Generator In-Situ Inspections"; GE Power Syst, GER 3954B (04/03) http://www.gepower.com/prod_serv/products/tech_docs/en/downloads/ger3954b.pdf.
59. P. Bagley, R. Roney, R. Hatley, K. Hatley, "Ultrasonic miniature air gap inspection crawler", Patent EP 01772949, General Electric Company, 9.10.2006.
60. SINTEF, "Robot That Climbs In The Pipe", ScienceDaily, 27 June 2008. http://www.sciencedaily.com/releases/2008/06/080624123559.htm.
61. H. Choset, "Modular Serpentine Robot Locomotion – Gaits, Locomotions and Behaviour", Info-page of Biorobotics Lab, Carnegie Mellon University, 2005, http://www.cs.cmu.edu/~biorobotics/projects/modsnake/gaits/gaits.html.
62. S. Roh and H. R. Choi, "Differential-drive in-pipe robot for moving inside urban gas pipelines", IEEE Transactions on Robotics, vol. 21, no. 1, pp. 1–17, Feb. 2005.

On poles or cables

63. K. Toussaint, N. Pouliot, S. Montambault, "Transmission line maintenance robots capable of crossing obstacles: State-of-the-art review and challenges ahead", Journal of Field Robotics Vol. 26-5, May 2009, 477-499.

64. R. Aracil, R. Saltaren, O. Reinoso, "A Climbing Parallel Robot. A Robot to Climb Along Tubular and Metallic Structures", IEEE Robotics & Automation Magazine, Volume 13, Issue 1, March 2006.

65. J. Maempel, E. Andrada, H. Witte, "INSPIRAT – towards a biologically inspired climbing robot for the inspection of linear structures", Proc. of the 11th International Conference on Climbing and Walking Robots and the Support Technologies for Mobile Machines (CLAWAR), September 2008.

66. M. Tavakoli, L. Marques, A. de Almeida, "Self Calibration of Step-by-Step Based Climbing Robots", The 2009 IEEE/RSJ International Conference on Intelligent Robots and Systems, October 11-15, 2009.

67. M. Abderrahim, C. Balaguer, A. Giménez, J. Pastor, V. Padrón, "ROMA: A Climbing Robot for Inspection Operations", International Conference on Robotics & Automation, ICRA'99, Detroit, USA, May, 1999.

68. J. Fauroux, J. Morillon, "Design of a climbing robot for cylindro-conic poles based on rolling self-locking", Proc. of the 12th International Conference on Climbing and Walking Robots and the Support Technologies for Mobile Machines (CLAWAR), 2009.

69. A. Sadeghi, H. Moradi, M. Ahmadabadi, "Analysis, simulation and implementation of a human inspired pole climbing robot", Proc. of The 11th International Conference on Climbing and Walking Robots and the Support Technologies for Mobile Machines (CLAWAR), Sept. 2008.

70. T. Sattar, H.Rodriguez, B. Bridge, "Climbing ring robot for inspection of offshore wind turbines", Industrial Robot: An International Journal, Vol. 36/4, 326-330, 2009.

Form fit or penetration

71. H. Amano, K. Osuka and T. Tarn, "Development of Vertically Moving Robot with Gripping Handrails for Fire Fighting", IEEE/RSJ International Conference on Intelligent Robots and Systems, Oct. 2001.

72. M. Bell, D. Balkcom, "A toy climbing robot". IEEE International Conference on Robotics and Automation (ICRA 2006), May 2006.

73. S. Kim, A. Asbeck, M. Cutkosky, W. Provancher, "SpinybotII: Climbing Hard Walls with Compliant Microspines", Int. Conf. On Advanced Robotics, 2005.

Climbing robots with pneumatic adhesion

74. DIEES - Università di Catania – Robotica, "Spiderbot II - A climbing robot for industrial inspection", http://www.robotic.diees.unict.it/robots/spiderbot2/spiderbot2.htm

75. C. Balaguer, A. Giménez, M. Abderrahim, "ROMA Robots for Inspection of Steel Based Infrastructures", Industrial Robot, Vol. 29/3, pp.246-251, 2002.

76. W. Brockmann, S. Albrecht, D. Borrmann, J. Elseberg, "Dexterous Energy-Autarkic Climbing Robot", Proc. of The 11th International Conference on Climbing and Walking Robots and the Support Technologies for Mobile Machines (CLAWAR), September 2008.
77. H. Zhang, W. Wang, J. Zhang, "A Novel Passive Adhesion Principle and Application for an Inspired Climbing Caterpillar Robot", Proceedings of the 2009 IEEE International Conference on Mechatronics, Málaga, Spain, April 2009.
78. C. Hillenbrand, D. Schmidt, K. Berns, "CROMSCI – A Climbing Robot With Multiple Sucking Chambers for Inspection Tasks", Proc. of The 11th International Conference on Climbing and Walking Robots and the Support Technologies for Mobile Machines (CLAWAR), September 2008.
79. DIEES - Università di Catania – Robotica, "Clawar Competitions 2000-2004 – Paris 2002 – Alicia 1", http://www.robotic.diees.unict.it/competitions/clawar/clawar.htm,, also a detailed publication exists but is difficult to access: D. Caltabiano, D. Longo, G. Muscato, M. Prestifilippo, G. Spampinato, "Learn to build Robots via Robotic competitions: the Experience of the University of Catania", Clawar/Euron Workshop ELH04, Vienna Austria, Dec. 2-4 2004.
80. D. Longo, G. Muscato, "A modular approach for the design of the Alicia3 climbing robot for industrial inspection", Industrial Robot: An International Journal, Vol. 31, Issue 2, pp. 148–158, 2004.
81. F. Bonaccorso, C. Bruno, D. Longo, G. Muscato, "Structure and model identification of a vortex-based suction cup", Proc. of The 11th International Conference on Climbing and Walking Robots and the Support Technologies for Mobile Machines (CLAWAR), September 2008.
82. J. Xiao, A. Sadegh, M. Elliott, A. Calle, A. Persad, H. Chiu, "Design of Mobile Robots with Wall Climbing Capability", Proceedings of the 2005 IEEE/ASME International Conference on Advanced Intelligent Mechatronics Monterey, California, USA, 24-28 July, 2005.

Climbing robots with other adhesion principles

83. S. Kim, M. Spenko, S. Trujillo, B. Heyneman, D. Santos, M. Cutkosky, "Smooth Vertical Surface Climbing With Directional Adhesion", IEEE Transactions on Robotics, Vol. 24/1, February 2008.
84. M. Murphy, M. Sitti, "Waalbot: Agile Climbing with Synthetic Fibrillar Dry Adhesives", Proc. of the 2009 IEEE International Conference on Robotics and Automation (ICRA), Kobe, Japan, May 12-17, 2009.
85. H. Tsukagoshi, A.Kitagawa, "Gel-Type Sticky Mobile Inspector to Traverse on the Rugged Wall and Ceiling,", Proc. of the 2009 IEEE International Conference on Robotics and Automation (ICRA), Kobe, Japan, May 12-17, 2009.
86. H. Prahlad, R. Pelrine, S. Stanford, J. Marlow, R. Kornbluh, "Electroadhesive Robots—Wall Climbing Robots Enabled by a Novel, Robust, and Electrically Controllable Adhesion Technology", 2008 IEEE International Conference onRobotics and Automation Pasadena, CA, USA, May 19-23, 2008.
87. O. Unver, M. Sitti, "Tankbot: A Miniature, Peeling Based Climber on Rough and Smooth Surfaces", Proc. of the 2009 IEEE International Conference on Robotics and Automation (ICRA), Kobe, Japan, May 12-17, 2009.

88. M. Greuter, G. Shah, G. Caprari, F. Tache, R. Siegwart, M. Sitti, "Toward Micro Wall-Climbing Robots Using Biomimetic Fibrillar Adhesives", Proc. of The 3rd International Symposium on Autonomous Minirobots for Research and Edutainment (AMIRE), 2005.

Other mobile robots, hybrids and manipulators

89. M. Lauria, Y. Piguet, R. Siegwart, "Octopus - An Autonomous Wheeled Climbing Robot", Proc. of The 5th Int. Conference on Climbing and Walking Robots (CLAWAR), 2002.

90. N. Elkmann, M. Lucke, T. Krüger, T. Stürze, "Kinematics and Sensor and Control Systems of the Fully Automated Façade Cleaning Robot SIRIUSc for Fraunhofer Headquarters in Munich", Proc. of The 6th International Conference on Field and Service Robotics (FSR), 2007.

91. T. Akinfiev, M. Armada and S. Nabulsi, "Climbing cleaning robot for vertical surfaces", Industrial Robot: An International Journal, Vol. 36/4 (2009) 352–357

92. R. Moser, N. Hugi, M. Bernhard, P. Isler, S. Honold, J. Erni, "Vorrichtung zur Inspektion eines Spaltes", Patent EP 2 110 678 A1, Alstom Technology Ltd., 14.4.2008.

93. D. Schafroth, S. Bouabdallah, C. Bermes, R. Siegwart, "From the Test Benches to the First Prototype of the muFly Micro Helicopter", Journal of Intelligent and Robotic Systems, 2008.

94. R. Alves, C. Ruella, M. Cabrera, L. Fermín, J. Cappelletto, M. Díaz, J. Grieco, G. Fernandez-Lopez, "communication system for the underwater platform poseibot", Proc. of The 12th International Conference on Climbing and Walking Robots and the Support Technologies for Mobile Machines (CLAWAR), 2009.

95. T. Sattar, H. Rodriguez, J. Shang, "Amphibious Inspection Robot", Proc. of The 11th International Conference on Climbing and Walking Robots and the Support Technologies for Mobile Machines (CLAWAR), September 2008.

96. G. Hélie, A. Lorotte, F. Conti, M. Lauria, J. Koller, "Tribolo – an onmnidirectional robot", Info on the K-Team –website, http://www.innowebtive.com/kteam/boards/kameleon/tribolo.html.

Overviews, classifications other PHD-theses

Overviews + classifications

97. M. Yim, "A Taxonomy of Locomotion", Chapter 3 of the PHD Thesis "Locomotion with a unit-modular reconfigurable robot", at the department of mechanical engineering of Stanford University, 1994.

98. D. Longo, G. Muscato, "Adhesion Techniques for Climbing Robots: State of the Art and Experimental Considerations", Plenary Talk at the 11th International Conference on Climbing and Walking Robots and the Support Technologies for Mobile Machines (CLAWAR), September 2008.

99. C. Balaguer, G. Virk, M. Armada, "Robot Applications Against Gravity", IEEE Robotics & Automation Magazine, March 2006.

100. H.Schempf, "In-pipe-assessment robot platforms - phase I - state-of-the-art review," Carnergie Mellon University, Pittsburgh, USA, Report to National Energy Technology Laboratory REP-GOV-DOE-20041102, Nov. 2004.

101. Z. Pan, Z. Zhu, "Miniature pipe robots", Industrial Robot: An International Journal, Volume 30, Number 6, 2003.

102. R. Siegwart, I. Nourbakhsh, "Locomotion Concepts", Chapter in the Book "Introduction to Autonomous Mobile Robots", Bradford Book, MIT Press, 2004. Short version in PPT-slides available at http://www.asl.ethz.ch/education/master/mobile_robotics/2_-_Locomotion_Concepts.pdf.

103. V. Hirschmann, J. Breguet, T. Cimprich, C. Groux, F. Rochat, "Système de locomotion pour micro-robots inspecteurs de turbines", Semester work at Laboratoire de systemes robotiques, EPFL, SS2007.

Other Master and PHD-theses with strong relation to this work

104. J. Catalá, E. Zwicker, W. Fischer, R. Siegwart, " Concept and Standardization of Probeholders used in robotic NDT inspection", Master Thesis at Autonomous Systems Lab at ETH Zürich, WS 2008/2009.

105. F. Tâche, "Robot Locomotion and Localization on 3D Complex Shaped Structures", PHD-Thesis at Autonomous Systems Lab, ETH Zürich, 2010.

106. T. Thueer, "Mobility evaluation of wheeled all-terrain robots. Metrics and application", PHD Thesis at Autonomous Systems Lab, ETH Zürich, 2009.

107. G. Caprari, "Autonomous Micro-Robots: Applications and Limitations", PHD at Ecole Polytechnique Federale de Lausanne (EPFL), 2003.

108. D. Longo, "Climbing robots: applications, design methodologies, control, experimental results", PHD Thesis at Università degli studi di Catania, 2003.

Websites where pictures were taken from

109. F. Mondada, F. Rochat, P. Schöneich, "Inspection Systems", Website of our partners at Laboratoire de Systemes Robotiques at EPFL, http://mobots.epfl.ch/inspection-systems.html

110. ALSTOM-Power, "GT24 and GT26 Gas Turbines – Features", Information from company-website, http://www.power.alstom.com/home/new_plants/gas/products/gas_turbines/gt24_gt26/features/39995.EN.php?languageId=EN&dir=/home/new_plants/gas/products/gas_turbines/gt24_gt26/features

111. R. Baake (Bundesumweltministerium), "Energieeffizienz und Erneuerbare - entscheidend für die zukünftige Energieversorgung", Eröffnungsansprache für die 7. Passivhaustagung, 25.11.2002, http://www.passivhaustagung.de/siebte/Umwelt.html

112. Blue-Water-Power – Energie aus Trink- und Abwasser, "Peltonturbine", Information from company-website, http://www.blue-water-power.ch/peltonturbine.html

113. E.ON Wasserkraft GmbH, "Wasserrad und Hightech-Turbine - Technisch das gleiche Prinzip – Turbinenarten → Kaplan-Turbine", Information from the company website, http://www.eon_wasserkraft.com/pages/ewk_de/Energiefakten/Regenerative_Energie/Funktionsweise/Turbinenearten_-_Fallhoehe_entscheidet/index.htm#

Die VDM Verlagsservicegesellschaft sucht für wissenschaftliche Verlage abgeschlossene und herausragende

Dissertationen, Habilitationen, Diplomarbeiten, Master Theses, Magisterarbeiten usw.

für die kostenlose Publikation als Fachbuch.

Sie verfügen über eine Arbeit, die hohen inhaltlichen und formalen Ansprüchen genügt, und haben Interesse an einer honorarvergüteten Publikation?

Dann senden Sie bitte erste Informationen über sich und Ihre Arbeit per Email an *info@vdm-vsg.de*.

Sie erhalten kurzfristig unser Feedback!

VDM Verlagsservicegesellschaft mbH
Dudweiler Landstr. 99
D - 66123 Saarbrücken

Telefon +49 681 3720 174
Fax   +49 681 3720 1749

www.vdm-vsg.de

Die VDM Verlagsservicegesellschaft mbH vertritt

Printed by Books on Demand GmbH, Norderstedt / Germany